AF308305

Brahim HERROU
Oumaima NACIRI

Dashboards measure the performance of a supply chain

Brahim HERROU
Oumaima NACIRI

Dashboards measure the performance of a supply chain

control of a supply chain by designing dashboards

ScienciaScripts

Imprint

Any brand names and product names mentioned in this book are subject to trademark, brand or patent protection and are trademarks or registered trademarks of their respective holders. The use of brand names, product names, common names, trade names, product descriptions etc. even without a particular marking in this work is in no way to be construed to mean that such names may be regarded as unrestricted in respect of trademark and brand protection legislation and could thus be used by anyone.

Cover image: www.ingimage.com

This book is a translation from the original published under ISBN 978-613-8-40515-3.

Publisher:
Sciencia Scripts
is a trademark of
Dodo Books Indian Ocean Ltd. and OmniScriptum S.R.L publishing group

120 High Road, East Finchley, London, N2 9ED, United Kingdom
Str. Armeneasca 28/1, office 1, Chisinau MD-2012, Republic of Moldova, Europe
Printed at: see last page
ISBN: 978-620-5-94176-8

Contents

Resume

In the industrial era, large companies could establish their position and ensure their success by incorporating technical advances into production equipment that allowed for mass production of standard products. The context was stable and competition was particularly weak and limited to firms from the same nation for local markets. Today, competition is no longer industrial but informational.

It is in this context of globalisation and therefore international competition that a growing number of companies and researchers recognise the benefits and importance of supply chain management in the search for business performance.

This is why they have been forced to review their management methods. Traditional management techniques have revealed their limits and have shown that financial indicators alone cannot ensure management.

In the framework of this work, we bring our reflexion to the problem of designing a performance management tool for a supply chain through the concept of the dashboard. This contribution takes the form of a robust and hybrid methodology, combining all the elements required to establish a dashboard and a system of relevant performance indicators.

Our methodology consists of four phases. The first phase consists of studying the field of study by determining the external environment and the information needed by the company to help design a dashboard. The second is to analyse the supply chain and model it via the processes. The third is to design the dashboards by process using a structured approach that is broken down into four steps. This approach integrates the result of an empirical study by questionnaire which provides information on the important performance indicators used in Moroccan industries. The last phase aims at developing a global dashboard of the supply chain.

Keywords: Supply chain, dashboard, performance indicators, methodology, questionnaire.

Abstract

At the time of the industrial era, large companies can establish their position and ensure success by incorporating technological advances in production equipment for manufacturing mass-standard products. The context was stable, competition was particularly weak and was limited to nation firms to local markets. Today, competition is no longer industrial but informational.

It is in this context of globalization and international competition so that a growing number of companies and researchers recognize the benefits and importance of control of the supply chain in the search business performance.

That is why they have been forced to revise their management. Classic management techniques have shown their limitations and have mostly shown the inadequacy of financial indicators alone able to ensure piloting.

As part of this work, we bring our thinking to the problem of designing a performance management tool for supply chain by the concept of Dashboard. This contribution takes the form of a robust and hybrid methodology that combines all the essential elements for the establishment of a Dashboard and a relevant performance indicator system.

Our methodology is divided into four phases. The first is to study the field of study by determining the external environment and the necessary company information helping to design a Dashboard. The second is to analyze the supply chain and modeling through process. The third is devoted to the design of dashboards process by using a structured approach which breaks down into four stages. This approach incorporates the results of an empirical study questionnaire which provides information on important performance indicators in Moroccan industries. And the last phase, aims to develop a comprehensive table edge supply chain.

Keywords: Supply Chain, dashboard, performance indicators, methodology, survey.

General introduction

In a changing economic context, the main challenge for Moroccan companies is to improve their competitiveness on a regional and international scale. In this perspective, companies have focused mainly on the criteria of industrial performance, namely cost, quality, delivery time and service, in order to meet the increased requirements of the final customer.

To meet this challenge, Moroccan companies aim to reorganise themselves to offer differentiated products while improving their reactivity and flexibility to achieve better performance of their supply chains.

Moreover, efficient management of the supply chain is inseparable from performance management at all times. Indeed, the supply chain is an essential component of the company's strategic approach in the quest for competitive advantage.

To support its strategy and perform all these functions, companies must then seek to define measures that contribute to the achievement of its strategic objectives, and on which it can act. Therefore, a well-designed management dashboard with well-chosen indicators is essential.

This theme of dashboard design tends to occupy a predominant place that is fairly representative of the unsatisfied expectations of companies in terms of management. Until the last few decades, the question of steering assistance was less present, when the context was stable and competition was particularly weak, and the best strategy was still to seek a continuous increase in productivity and a reduction in costs. The scoreboards of that time, limited to exclusively financial and productivity measures, were quite appropriate.

Today, the context has changed considerably, and in order to guarantee a real return on invested capital, it is necessary to elaborate well-considered strategies. The monitoring of financial measures is not sufficient, the strategy loop is too slow and does not allow to react in time.

In this work, we are interested in the management of the supply chain of a company by the design of a global dashboard. The aim is to define a robust and generic methodology for designing a dashboard measuring the performance of the supply chain.

Our methodology, articulated in four phases, is a contribution that couples :

- 1 top-down approach that is effective in ensuring implementation

of the strategy, as defined in the ruling circles;

- And the buttom - up approach involves the process actors and takes into account the operational constraints. It is based on the different process malfunctions that affect the whole supply chain, either in terms of cost, time or quality.

Knowing that the choice of really relevant indicators is the key to success of any management project, we will carry out an empirical study by questionnaire to identify the main indicators considered important for companies. We will then propose an effective method for selecting and constructing real indicators that contribute to decision making.

Our methodology will be validated by an industrial application in a Moroccan small and medium enterprise (SME).

The work of our thesis is divided into three parts:

- The **first part** includes two chapters. **Chapter 1** will introduce the concepts of logistics, the supply chain and supply chain management. **Chapter 2** will focus on the measurement of the performance of a supply chain and the integration of risk management into its performance measurement system. The concept of indicators has been well addressed. In fact, the implementation procedures of a system of indicators will be summarised in order to choose those which will be recommended in our study.

-**Part Two** consists of two chapters. In **Chapter 3,** we will define a dashboard in order to better understand the services it can provide to a manager and we will introduce the main approaches and practical methodologies for its implementation. **Chapter 4** is reserved for the description of our methodology which aims at steering the operational and global supply chain through a dashboard. In this chapter, we propose an original methodology based on four phases. For each of these phases, we propose the steps leading to the achievement of their objectives as well as the tools used. In order to build a bank of performance indicators adapted to the industrial sector, we will carry out an empirical study on the main indicators considered important for the industries that we will carefully detail.

-**The third part** consists of three chapters that will be reserved for the validation of our methodology. **Chapter 5** will be devoted to the application of the first two phases of our methodology to a small and medium-sized enterprise (SME). **Chapter 6** will focus on the application of the third phase of our methodology. This phase consists of designing dashboards for each process in the chain via a general approach based on the following steps:

>   Step 1: Process analysis ;

>   Step 2: Risk identification and target setting;

>   Step 3: Choice of indicators ;

>   Step 4: Design a process dashboard.

Chapter 7 will focus on the application of phase 4 of our methodology to build a global supply chain scorecard for our case study.

Finally, we will present the conclusions of our work and we will propose some research perspectives from industrial and academic points of view.

Part 1

State of the art and positioning

The concepts of supply chains, supply chain management and performance measurement through indicators are presented in this section. The most comprehensive review possible is presented in order to enable non-specialists to understand the issues considered in the following chapters. Our literature review shows the paucity of scientific publications on the measurement of supply chain performance in the Moroccan industrial context.

Chapter 1: General Supply Chain Issues

1.1 Introduction :

This chapter aims to give some generalities and definitions used by the scientific community working in the field of supply chain. This is done by addressing :

> the definition of logistics ;

> the concepts of supply chain and its management;

> the model opts for the modelling of supply chain processes (the SCOR model).

At the end of this chapter, we will present a synthesis and the positioning of our work in relation to the literature.

1.2 Definition of logistics

The word "logistics" appeared in France in the 18th century, when the problems of supporting military strategy (replenishment of weapons, ammunition, food, horses, uniforms, shoes, etc.) were no longer neglected. This term then spread, especially in the industrial world, to refer mainly to the handling and transport of goods.

Until the 1970s, logistics had little importance in the management of companies, considered as a secondary function, limited to execution tasks in warehouses and on shipping docks. However, logistics is now understood as an operational link between the different activities of the company, ensuring the coherence and reliability of the material flows, with a view to the quality of the service to the customers while allowing the optimisation of resources and the reduction of costs.

In the mid-1990s, logistics became a global or even worldwide function of physical flow management in a complete vision of the Customer/Supplier chain, and truly constitutes a new discipline of company management. Global logistics" thus represents all the activities, internal or external to the company, that add value to products and provide services to customers (Courty, 2003).

Today, logistics is a function, more or less mature in companies, but also a discipline and an object of study for researchers. More precisely, the most

The most recent development in logistics is supply chain and supply chain management (Belin-Munier, 2014).

1.3 Definitions of a supply chain

After much debate, the concept of a supply chain is now commonly understood as a network of facilities that provides the functions of procurement of raw materials or semi-finished goods, transportation and processing of these materials into components, semi-finished goods and then finished goods, and finally storage and distribution of finished goods to customers (Lee and Billington, 1993).

One understanding of the concept of supply chains considers the latter as a set of successive customer/supplier relationships integrating, for each entity, the activities of supply, production and distribution (Tayur et al, 1999) (Stadtler, 2000). This proposal completes the previous definition by focusing the supply chain on the relationships between the actors that compose it.

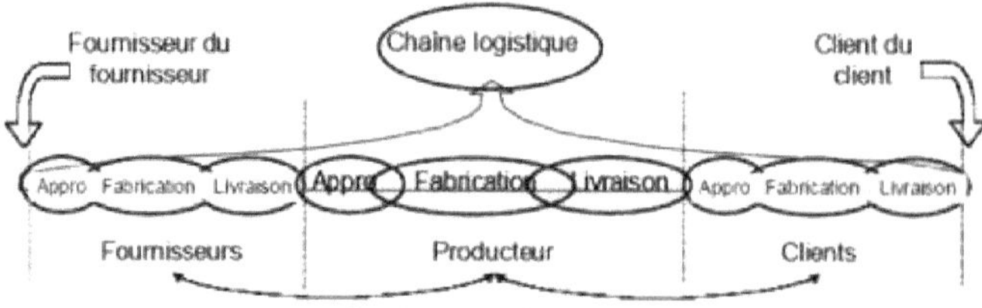

Figure 1.1 Customer/supplier relationships in a supply chain and functions present

We can say that a supply chain consists of the entire path of physical, financial and informational flows, from the first suppliers to the ultimate customers, the consumers (Figure 1.1). A supply chain therefore appears, at first glance, to be very large, for two main reasons. Firstly, there is always a supplier to the supplier. Secondly, it is still very difficult to know where the consumption of a product ends when the recycling process is included in the reflection... Thus presented, the supply chain seems to be a totally disproportionate entity.

However, it is important to understand that there are smaller chains. Thus, the various sites of a large company or its main functions can each be the customer or supplier of another entity, and thus also represent links in logistics chains.

Poirier and Reiter (2001) simply state that a supply chain is the system by which companies bring their products and services to their customers. The supply chain is here clearly associated with the firm (or network of firms) and therefore with the processes and resources that drive it.

Francois (2005) has summarised a set of definitions in a table. He then points out that these definitions can however be categorised according to their main orientation, a supply chain is a succession of Customer/Supplier relationships, value creation activities or functions (supply, transformation and distribution processes).

In the supply chain, manufacturers, intermediaries, traders, transport companies, suppliers and official bodies collaborate to provide products at the right time and of the right quality (Campuzano, Mula, and Peidro, 2010, Langroodi, 2016). In terms of interdependence between the different entities of a supply chain, Lavastre et al (2012) equate supply chains to systems composed of interdependent, evolving elements, in which disruption in one element can negatively influence the performance of the entire supply chain.

1.3.1 Supply chain models

The best known models that incorporate the representation of the supply chain concept are the SCOR model (Figure 1.2) and the one proposed by (Kearney, 1994) (Figure 1.3).

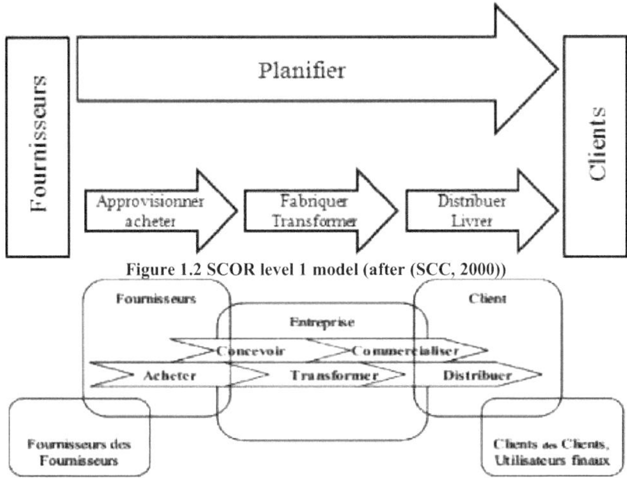

Figure 1.2 SCOR level 1 model (after (SCC, 2000))

Figure 1.3 Supply chain model from (Kearney, 1994)

Both models emphasise the key functions of the physical flow management processes of procurement (including purchasing), production (in the broad sense) and distribution. The SCOR model adds the fundamental notion of planning and thus makes the link with the management component. (Kearney, 1994) emphasises the basic interaction that remains between the physical flow and demand management on the one hand (selling, marketing) and the creative process on the other (designing, innovating). It would have been possible to add also a support activity aiming at maintaining the products and services transmitted to the customers. (Dupont, 2003) calls this activity 'logistical support'.

1.3.2 Supply chain approaches

Pichot (2003) distinguishes three main categories of definition:

J The founding element of the supply chain is the company (internal supply chain) In this context, a company is a succession of functions, which can be assimilated to a supply chain of functions or internal supply chain;

J The supply chain extends from the supplier to the customer (integrated supply chain); The supply chain can also be defined in a more functional way: a supply chain is a network of facilities that perform the functions of supplying raw materials, transforming these raw materials into components and then into finished products, and distributing the finished products to the customer;

J The supply chain includes the supplier's supplier and the customer's customer (Collaborative Supply Chain): the most general and extended definition defines the supply chain as a system whose components are suppliers, production plants, distribution services, and customers linked together by upstream to downstream material flows and information flows in the other direction. This definition makes it possible to extend the supply chain beyond the boundaries of the company and the supplier/company/customer trio, and thus to define a supply chain from suppliers of suppliers to customers of customers (Figure 1.4).

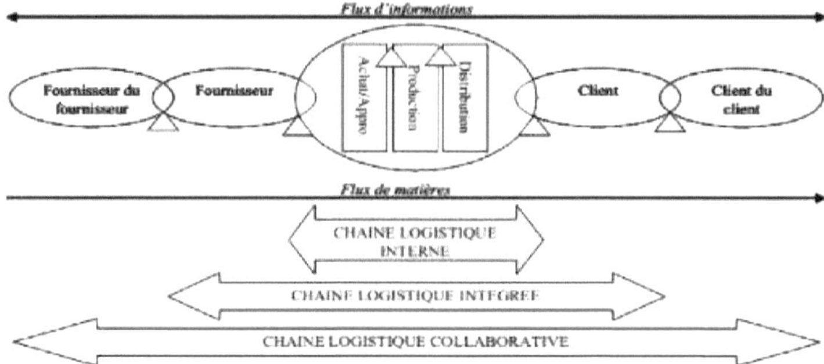

Figure 1.4 Different supply chain approaches

1.4 Flow management in the supply chain

We detail here the three flows that cross a supply chain: information, physical and financial. These three flows can be derived from the rules stipulated in the partnership contract. Indeed, contracts define the relations between each company in the supply chain, providing for penalties in the event of late delivery by a supplier or stock shortage, determining who manages transport and stocks between two "links" in the chain, etc.

1.4.1 Information flow

The information flow represents all the transfers or exchanges of data between the different actors in the supply chain. This is primarily commercial information, in particular orders placed between customers and suppliers. An order generally includes the product reference, the quantity ordered, the desired delivery date and the price, if any, negotiated at the time of sale. Other elements can be added to this list: the list of options desired for the product, the delivery frequency if necessary, ... But companies also exchange more technical information: physical parameters of the product, operating ranges, production and possibly transport capacities, stock level monitoring information. The latter is increasingly requested by customers who wish to know the state of progress of their product. More generally, the principle of traceability translates into an increased right of scrutiny by the customer towards the supplier (Dupuy et al, 2004).

The flow of information is becoming faster and faster thanks to advances in ICT. The development of information flows within the supply chain is limited by the need for confidentiality between actors. Moreover, the problem of the quality of the data conveyed remains, and there is a risk that decisions will be based on erroneous or simply peremptory data.

1.4.2 The physical flow

The physical flow is constituted by the movement of goods transported and transformed from raw materials to finished products through the various stages of semi-finished products. It justifies the organisation of a logistics network, i.e. the different sites with their production resources, the means of transport to link these sites and the storage space needed to compensate for contingencies and to act as a buffer between two successive activities. In short, the physical flow results from the implementation of the various handling and processing activities of the products whatever their state. The physical flow is generally considered to be the slowest of the three flows.

1.4.3 Financial flow

The financial flow concerns all the pecuniary management of companies: sales of products, purchases of components or raw materials, but also production tools, various equipment, warehouse rental, ... and of course the employees' salaries. The financial flow is generally managed centrally in the company's financial or accounting department, but is linked to the production function by the purchasing and sales departments. In the long term, it also corresponds to heavy investments such as the construction of new buildings and production lines.

1.5 Supply chain functions

Generally, the functions of a supply chain range from the purchase of raw materials, through production, storage and distribution, to the sale of finished products.

1.5.1 Procurement

It is the most upstream function in the supply chain. The materials and components supplied constitute 60% to 70% of the costs of the products manufactured (Ouzizi, 2005) in a majority of companies. Reducing procurement costs helps to reduce the costs of finished products, and thus to increase financial margins. The delivery time of suppliers and the reliability of distribution have a greater influence than production time on the stock level and service quality of each manufacturer (Harmon, 1992).

1.5.2 Production

The production function is at the heart of the supply chain, it is the skills that the company has to manufacture, develop or transform raw materials into products or services. It gives the capacity of the supply chain to produce and thus gives an indication of its responsiveness to fluctuating market demands. If factories have been built with a large, sometimes excessive, production capacity, then they can be reactive to demand in the presence of extra quantities to be made, this environment has the advantage of being available for customers in case of urgent demands, but on the other hand part of the production capacity can remain idle which generates additional costs and expenses. On the other hand, if the production capacity is limited, the supply chain has difficulty in being very reactive and therefore may lose market share as it is not able to respond favourably to certain demands. A balance must therefore be found between responsiveness and costs.

1.5.3 Storage

Storage includes all quantities stored throughout the process, starting with raw material stock, component stock, work-in-progress stock and finally finished product stock. The stocks are therefore shared between the different actors: suppliers, producers and distributors. Here too, the question arises of the balance to be struck between greater reactivity and cost reduction. It is obvious that the more stocks you have, the more responsive the supply chain is to fluctuations in market demand. However, having stocks generates costs and risks, especially in the case of perishable products or products whose speed of innovation is such that a new range of the same product put on the market by a competitor can make the quantities of this product in stock obsolete and thus a significant loss (Mouloua, 2007).

1.5.4 Distribution and transport

The transport function is involved throughout the chain, transporting raw materials, transporting components between factories, transporting components to storage centres or to distribution centres, and delivering finished products to customers. The relationship between the reactivity of the chain and its efficiency can also be seen in the choice of transport mode. The fastest modes of transport, e.g. aeroplanes, are very expensive, but allow for a very quick response to unpredictable demands. Rail or truck transport is more cost effective but not as fast. All partners can choose to combine these modes of transport and adapt them to certain situations according to the importance of the demand and the total gain generated.

1.5.5 Selling

The sales function is the ultimate function in a supply chain, its efficiency depends on the performance of the upstream functions. If the previous steps have been well optimised, then the sales staff will have an easier time of it, as they will be able to offer more competitive prices than the competition, otherwise the margins will be very small and the profits not very high, or even losses will be generated.

1.6 . Structuring decisions in a supply chain

A decision can be defined as the problem of giving a value to an unknown variable, the knowledge of which allows the decision-maker to escape from a situation of judgment or uncertainty (Ouzizi, 2005). The design of a supply chain requires a set of decisions to be made. This set of decisions can be considered on three hierarchical levels: strategic decisions, tactical decisions, and operational decisions.

1.6.1.   The strategic level

Strategic decisions (time horizon: month to year) involve the definition of investment policies, management and design of the logistics network (Steadtler and Kilger, 2001), (Dudek and Stadtler, 2005) or the reconfiguration of an existing network (Pirard, 2005). They involve significant financial investments (HolaBa et al, 2016).

These decisions are guided by the establishment of financial and commercial objectives and are aimed at defining the profile of partners, the location of infrastructure, the capacity required per logistic service unit.

1.6.2.   The tactical level

Once the strategic decisions have set the guidelines for the configuration of the partner network, the tactical decisions (horizon: week to month) are concerned with the implementation of medium-term plans in order to programme the activities to be carried out within each company. These include the procurement, production / stock management and distribution processes, and are concerned with rationalising them as best as possible with regard to the objectives defined by the company strategy.

1.6.3. The operational level

Decisions at the operational level (time horizon: day to week) consist of actions planned at the tactical level. They result from a decomposition of tactical decisions into detailed operations (HolaBa et al, 2016). They concern the scheduling of activities, the adjustment of plans according to the hazards and disturbances observed during monitoring and performance measurements (Figure 1.5).

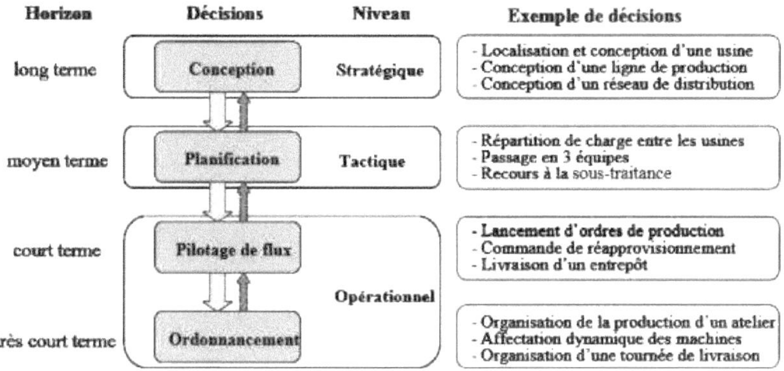

Figure 1.5 Decision-making levels in a supply chain

1.7 Supply Chain Management (SCM)

Supply chain management refers to the so-called SCM (Supply Chain Management) approaches.
Chain Management) in the Anglo-Saxon scientific literature. Chain management
Logistics is characterised by definitions as diverse as the disciplines and currents from which they originate.

1.7.1 The SCM approach

Humez (2008) has cited a set of definitions of supply chain management which (Okar, 2011) summarises in Table 1.1.

Table 1.1 Definitions of supply chain management (OKAR, 2011)

Author	Description
Stevens, 1989	The reason for managing supply chains is to synchronise the needs of customers with the flow of materials from suppliers in order to achieve a balance between the following conflicting objectives: To have a high quality of service, low stock and to produce products at low cost.
Christopher, 1992	SCM is the management of upstream and downstream relationships with suppliers and customers, to provide the supply chain with superior customer value at the lowest cost.
Cooper and Ellram, 1997	SCM is an integrative philosophy for managing the distribution flow from the supplier to the end customer.
Mentzer et al, 2001	SCM is about improving both efficiency (i.e. cost reduction) and effectiveness (i.e. customer service) in a strategic context (creating customer value and satisfaction through integrated supply chain management) to achieve competitive advantages that yield profit.
Gibson et al, 2005	SCM essentially integrates supply and demand

	management within companies.

To these definitions, we can add those of Simchi-Levi et al, (2003) who define SCM as "*a set of approaches used to effectively integrate suppliers, producers, distributors and retailers in order to ensure the production and distribution of finished products at the right time, in the right place, in the right quantity, meeting the requirements of the final customers, and at the lowest cost*".

For (Steadtler, 2005) SCM (Supply Chain Management) is defined as: "*the process of integrating organisational units along the supply chain and coordinating physical, informational and financial flows with the aim of satisfying the end customer and improving the competitiveness of the chain as a whole*".

These two definitions, without being exhaustive, address a main foundation of SCM, namely an integrative approach of all partners (from the supplier to the final customer) aiming at achieving local and global performance objectives. The same findings as (Regragui et al. 2013, Boudahri, 2013) state that SCM implies a synchronisation of all activities in the chain by integrating suppliers, manufacturers, warehouses, distributors, retailers and customers. The objective is to create added value for the customer and all SC actors.

According to (Ramanathan, 2014) the SCM organises and manages the entire process of supply network activities from suppliers through manufacturers, to retailers/wholesalers to the end user.

1.7.2 Approaches to supply chain management

Supply Chain Management can therefore be limited to the boundaries of a company, extend to its direct partners or integrate several companies in the chain. The length of the chain studied is therefore a measure of the scope of supply chain management. There are also three types of approach based on the different levels of supply chains (Pichot, 2006):

J **Management of internal supply chains**

This approach focuses on the operational efficiency within a company through the search for optimisation of physical flows and improvement of operational processes. Studies in internal supply chain management are mainly concerned with problems related to the supply, production or distribution of goods within a company.

J **Integrated supply chain management**

This approach focuses on the relationships between several sites of a company, even including some of the company's direct suppliers or customers. "*Supply Chain Management is the delivery of enhanced customer and economic value through synchronized management of the flow of physical goods and associated information from sourcing to consumption*" According to (La Londe, 1994).

J **Collaborative supply chain management**

This approach consists of working at the level of a company within the supply chain(s) to which it belongs. According to (Bowersox et al.1997) SCM is distinguished from SC (Supply Chain) by its reference to management tools and methods that ensure the optimisation of the whole chain. The SCM can be seen as the set of methods associated with the performance improvement approaches that have succeeded each other in recent years (Amrani Zougari, 2009), whose main objective is to improve industrial competitiveness by minimising costs, The main objective is to improve industrial competitiveness by minimising costs, ensuring the level of service required by the customer, efficiently allocating activities to production, distribution, transport and information actors, and ensuring that actors do not develop antagonistic local behaviours that undermine overall performance.

1.7.3 Supply chain processes

The definition of supply chain management invariably leads us to the notion of process, since supply chain management implies an organisation by process rather than by function.

According to ISO 9000 version 2000, a process uses resources and is managed in such a way as to enable the transformation of input elements into output elements. The supply chain is then often assimilated to a system composed of a set of processes that are strongly correlated with each other and we speak of the management of the supply chain by processes (Morley, 2002); this approach corresponds to the continuous improvement of the supply chain via the evolution of processes and their interfaces.

Many models relating to the characterisation or management of a supply chain are built around the identification of its processes. Among these works, one finds mainly audit-type approaches, analysis methodologies or supply chain diagnostic tools (France-Anne, 2007).

There are five models for identifying the main processes in supply chain management: the Gilmour (1999) and Cooper et al. (1997) models, the SCOR model (SCC, 2011), the ASLOG logistics guide (2006), and the EVALOG reference model (2006).

The three main models (Scor, Gilmour, Cooper) base the modelling of the supply chain on the processes that compose it. Their main objective is to provide a framework and representation of the supply chain in terms of process typology and content (Ibn El Farouk, 2013). These models have been used by a large number of researchers for various purposes; implementation of a supply chain project, elaboration of a performance benchmark, supply chain optimization, supply chain integration.

In our thesis, we limit our study to the most internationally recognised model for its application to supply chains, which is the SCOR-model (Supply Chain Operations Reference Model), whose characteristics and applications we will detail.

1.8 The SCOR (Supply Chain Operations Reference-model)

Given the level of complexity of today's supply chains, it is useful to identify and map the company's processes in the form of a model. Establishing a common language in the company is essential because it is a matter of semantic rules that structure the actors' discourse and determine the relationships between the words used (Wattky Crestan, 2006).

The SCOR model was developed as an international standard language by the Supply Chain Council (SCC). The SCC is a not-for-profit association established in 1996 by representatives of Advanced Manufacturing Research (AMR), Bayer, Compaq Computer, Pittiglio Rabin Todd and McGrath (PRTM), Procter and Gamble, Lockheed Martin, Nortel, Rockwell Semiconductor and Texas

Instruments. The SCC is intended to promote the SCOR model, launched in 1997.

SCOR assumes that any supply chain can be subdivided into five types of business processes (planning, procurement, manufacturing, delivery, returns), which are considered to be the core processes of any supply chain. The model regularly confronts them with the analysis of best practices, observed in the industry, with benchmarking studies as well as with different existing IT solutions.

1.8.1 Activities of the Supply Chain Council (SCC)

The SCC has two websites (http://www.supply-chain.org, http://www.supplychainworld.org) on which information about SCOR and other supply chain management information is available to members. The latest version of the SCOR model is also available to members. Its studies allow SCC members to compare the performance of their supply chain with that of other companies in and outside their industry.

In addition, SCC members can participate in technical committees, whose purpose is to create working groups for the further development of SCOR. There are six technical committees (planning, procurement, manufacturing, delivery, return and model integration). These committees are responsible for the content of each part of SCOR and for ensuring that the model remains coherent and integrated.

1.8.2 The main objectives of the Supply Chain Council

The SCC aims to create, improve, test and validate standard supply chain processes from supplier to supplier to customer. The SCOR model covers all types of businesses and is usable by all functions and departments in the supply chain. The SCC also seeks to identify key metrics to measure the performance of SCOR standard processes and to use these metrics in benchmarking studies of other companies. The information from the benchmarking studies is incorporated into the definition of best practices, recommended by the model, for each process element of the model. The SCC also lists the various software products available to address these best practices and allows each member of the SCC to contribute to the continuous improvement of the model and to the improvement of their own supply chain management.

The CSC finally seeks to promote the processes, incorporated in the model, as a valid standard for all types of industries and functions for supply chain management, thus reaching a wide user population.

1.8.3 The content of the SCOR model

It is a reference model that integrates the following three reference elements:

-    BPR: concerns the reorganisation of processes through process descriptions. It is about capturing the current state of the existing process and determining its desired state in the future by aligning it with the ideal process that is described by SCOR;

-Benchmarking: is the evaluation of an organisation's goods, services or practices by comparison with models that are recognised as reference standards. It is about measuring the operational performance of similar companies and making internal targets on the best of them. This also represents the basis for best practice models;

\-   Best practice analysis: concerns the knowledge or ways of doing things that lead to the desired result and that are exemplified to peers in order to share the experience that will allow collective improvement. It is the characterisation of the organisational and information systems by which best practice is achieved, as observed in the industry.

The SCOR model therefore includes a standard language of management processes, a framework of relationships between processes, standard parameters for measuring process performance and management practices that result in best practice.

This model is concerned with processes and procedures and not with the functions of the company. In other words, it focuses on the activities involved and not on the individuals or elements of the organisation that carry out these activities. However, the model does not cover functions such as marketing, after sales service, etc. SCOR is composed of several sections that are organised around five main management processes: planning, sourcing, manufacturing, delivery and return activities (Figure 1.6).

Figure 1.6 Five main processes of the SCOR model

Through a common set of definitions, the model can be used for a wide range of supply chain types, from simple to complex. As a result, heterogeneous industries can be linked to describe the depth and breadth of virtually any supply chain.

SCOR's standard definitions cover processes, terminology, performance indicators, best practices and, where necessary, a description of the information technology that supports the model. Technology is seen by SCOR as a tool that can help to improve the organisation of the company. However, it is the processes that are continually emphasised as they are the ones that drive the identification and prioritisation of opportunities and the improvement of overall process improvement, which are not necessarily linked to technology.

With regard to performance indicators, the model proposes a large number of indicators that are grouped under five performance attributes: *reliability* (measured by, among other things, the percentage of goods on the shelf, replenishment accuracy, on-time delivery performance, execution, etc.), efficiency (measured by, among other things, the percentage of goods on the shelf) and quality. "This is the case for *the following factors*: *the number of days* of *stock* in the warehouse (number of days of stock in the warehouse, etc.), the number *of days* of stock in the warehouse (number of days of stock in the warehouse, etc.), the number of days of stock in the warehouse (number of days of stock in the warehouse, etc.), the number of days of stock in the warehouse (number of days of stock in the warehouse, etc.), the number of days of stock in the warehouse (number of days of stock in the warehouse, etc.), the number *of days of stock in the warehouse, etc.), the number of days of stock in the warehouse, etc.), the number of days of stock in the warehouse,* etc.

The company can then focus and measure performance indicators according to its strategic, tactical and operational priorities. SCOR contains three distinct levels of detail but does not intend to impose its views on the methods that should be used in a business to manage the flow of information generated by the systems already in place. In order to link the company's strategy to its supply chain, SCOR is built using a top-down approach. The processes are described in sub-processes which are in turn broken down into activities.

The first level of the model concerns the different types of processes (planning, procurement, manufacturing, delivery, returns). This is the scope and content of the model applied in the company. The performance and competitiveness objectives are defined at this level.

The second level two concerns the configuration of the supply chain. Since three types of production are distinguished (make-to-stock, make-to-order and design-to-order), the corresponding procurement and delivery processes are also distinguished in this way.

The third level is the process elements. Here the sub-processes are broken down into different activities, allowing the company to define its ability to compete successfully in its market. This level consists of defining the different tasks required in the process, identifying the input and output data, identifying and analysing the corresponding best practices and measuring the system's ability to support the process in question.

The fourth level is not part of the SCOR model because it is about the implementation of processes and therefore procedures, systems, etc. that are specific to each company. The latter sets up here the specific practices of its Supply Chain Management, defining the practices to be implemented to develop competitive advantages and adapt to the varying conditions of its market.

1.8.4 The five main processes of the SCOR model

Like the performance indicators and best practices of the SCOR model, each of the five management processes of the model has a definition and is linked to a specific process category.

The planning process (Plan) is concerned with balancing all demand and supply to develop a plan of action that meets the needs of supply, production and delivery and returns. The process categories concerned are P1 (supply chain planning), P2 (procurement planning), P3 (production planning), P4 (delivery planning) and P5 (returns planning).

The Source process is concerned with the procurement of products and services to meet expected, planned or actual demand. The process categories concerned are S1 (Source Stocked Product), S2 (Make-to-Order) and S3 (Engineer-to-Order).

The Make process involves the transformation of supplied products into finished products to meet expected, planned or actual demand. The process categories are M1 (Make-to-Stock), M2 (Make-to-Order) and M3 (Engineer-to-Order).

The Deliver process is concerned with the dispatch of products and services to customers to meet expected, planned or actual demand. The process categories are as follows: D1 (Deliver Stocked Product), D2 (Deliver Make-to-Order), D3 (Deliver Engineer-to-Order) and D4 (Deliver Retail Product, where the customer goes to a depot or shop to take delivery of their product).

As for the return process, it concerns the activities of returns of products received from customers or products to be returned to suppliers. There are two categories of processes concerned: returns to suppliers (R1 - return of defective products, R2 - return of products to be repaired and R3 - return of excess products) and customer returns (R1 - return of defective products, R2 - return of products to be repaired and R3 - return of excess products).

Finally, there is a process, called **Enable**, which is a term that refers to the activities that lead to a desired state. Enable considers the five processes mentioned above and helps to establish and manage business rules, evaluate performance, manage data, inventory, capital, transportation, supply chain configuration and product/service compliance.

1.8.5 Interests and limitations :

The value of SCOR is reported in the work of (Wong and Wong, 2008), who state that companies adopting SCOR enjoy a standard format that facilitates communication and thus enables benchmarking. (Geary and Zonnenberg, 2000) cite a benchmarking study that shows that supply chains using the SCOR model have experienced significant financial gains and operational benefits According to (Wattky Crestan, 2006), one of the weaknesses of the SCOR model remains the absence of aggregation mechanisms that show how to combine indicators defined at a detailed level to inform those defined at the aggregate level. Moreover, each organisation selects from among the metrics proposed by SCOR the indicators that it considers useful for the management of its activities without guaranteeing the construction of relevant and coherent dashboards at the scale of the supply chain. According to (Samuel et al, 2004), although SCOR does provide a common framework and indicators, the approach is perceived by some as too rigid and in need of significant refinement and improvement to adapt to the increasing complexity of supply chains and the frequent changes that need to be managed.

1.9 Summary and positioning

In this context of multiple definitions, we propose to agree on a definition of the supply chain. Our work will be centred on the objective of measuring performance through a dashboard. This is why two notions seem interesting to us to deepen among the multiple definitions stated previously: the internal supply chain and the integrated supply chain:

-    The internal supply chain represents all the actors of the company involved from the purchase of raw materials from suppliers to the delivery of products to customers. The associated problems thus correspond to the control of information, physical and decisional flows along the internal chain. It is a transversal vision of the company's operation based on the processes that run through it;

-    The integrated supply chain places the company within a chain or network of actors extending from the first supplier to the final customer. In this context, it consists of the company, its customers and suppliers. The aim of the work on the integrated supply chain is to improve cooperation and coordination with actors outside the company.

We can say that the integrated supply chain includes the internal supply chain plus the supplier and the customer.

We will therefore use these two terms throughout the thesis: "supply chain", which should be understood in the sense of the company's internal supply chain, and "integrated supply chain", which defines the logistics network in which the company interacts with its customers and suppliers.

For the definition of supply chain management, we will adopt the following definition: ***Supply chain management is an integrative approach to agreeing on the planning and control of the physical flow between all the participants in the supply chain (suppliers, producers, distributors), from the raw material to the finished product, so that the goods are produced and distributed in the right quantity, at the right place and at the right time.***

In order to identify the main processes of supply chain management, we will choose the SCOR model for the following reasons:

>   A reference model of the supply chain;

>   An architecture based on the process approach ;

>   An architecture that includes a performance measurement component;

>   A strategic and operational modelling of the supply chain ;

>   Its use for a wide range of supply chain types, from simple to complex, and its success in internal supply chain management.

1.10 Conclusion

In this first chapter we have :

>   Clarifies the concepts of supply chain and supply chain management;

>   Specifies two terms that we will use throughout the brief:

■   The term "supply chain" should be understood in the sense of the company's internal supply chain;

■   The "integrated supply chain" defines the logistics network in which the company interacts with its customers and suppliers.

■   Presents the models to identify the main processes characterising the management of a supply chain, in this case the SCOR model.

Finally, we will look at measuring the performance of the supply chain. This will be the subject of the next chapter.

Chapter 2: Performance measurement in the supply chain

2.1 Introduction

After defining the concepts related to supply chains and their management in the previous chapter. In this chapter, we focus on the problems related to performance management in supply chains. We will first:

- Discuss the concept of performance measurement;
- Situate the concept of performance measurement in the supply chain;
- To clarify the concept of risk and risk management in the supply chain;
- Addressing the integration of risk management into the

logistics performance.

Then we will look at the indicators to best measure performance. We will see that these indicators must be structured in order to be meaningful to managers. Then, we will detail the methods for implementing a system of performance indicators (SIP) for a supply chain. Finally, we will present a bibliographic synthesis.

At the end, we will detail the approaches to implementing a SIP recommended in our methodology for efficient supply chain management.

2.2 Supply chain performance

2.2.1 Performance is a complex concept

Performance is a difficult concept to grasp. A quick search of the literature shows that there are many definitions of performance - each with its own definition - which contributes to the notion being a catch-all word with many meanings (Saulquin and Maupetit, 2004).

The difficulty in understanding this notion stems from its relative, subjective and even paradoxical nature. Performance has as many facets as observers, and for some it remains "a matter of perception" (Saulquin and Schier, 2005).

The notion of performance is also evolving. Taylor's notion of performance at the beginning of the twentieth century is quite different from Hollnagel's (2006) today. Taylor associates the performance of the firm with the division of labour, the scientific selection of workers, the improvement of their knowledge, fair payment, etc. For Hollnagel, the performance of the firm is the result of the division of labour. For Hollnagel, the performance of the firm is inscribed in its organisational resilience, i.e. its intrinsic capacity to recognise and adapt to changes, to aggressions and to return to a stable state (Hollnagel et al, 2006).

The evolving nature of the concept, its multiple facets, the difficulty of representing it, of describing it, thus highlight the complex nature of the concept of performance and the resulting difficulty of fully understanding it.

2.2.2 Logistics performance

Logistics performance is a multiple concept that must be understood in a transverse and global way, as the flows do not stop at the company's borders. Its translation is however not obvious in the face of the complexity of the logistics chain (SC Meter, 2015).

In his article, Morana (2012) considers that the measurement of logistics performance aims at continuous improvement, which leads to the conceptualisation and implementation of measurement systems that combine diagnosis and decision support.

Whatever the objectives pursued by companies, the ultimate goal of the supply chain is to meet customer demand at the lowest cost with the least impact on the environment. In this sense, supply chain performance is defined as the result of four key factors, namely **reliability, efficiency, responsiveness and environmental friendliness**, on which any supply chain manager must act in order to fulfil its mission (SC Meter, 2015) (Figure 2.1).

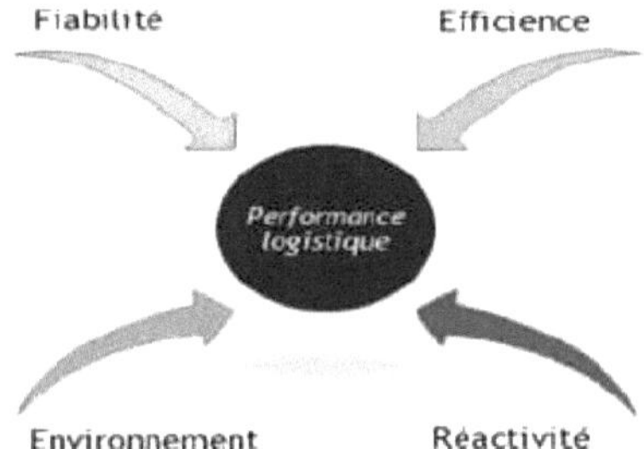

Figure 2.1 The levers of logistics performance

Logistics reliability: An organisation is said to be reliable when the probability of fulfilling its mission over a given period of time corresponds to that specified in the contract or specifications. For logistics, reliability means the ability to meet customer demand at a fixed service level.

Logistical efficiency: Efficiency is the ratio of effectiveness to cost. It refers to the achievement of an objective with the minimum of means possible.

Logistics responsiveness: Responsiveness is one of the essential properties of a company operating in a highly variable market context. It characterises its ability to quickly propose the most appropriate response to disturbances (of internal or external origin to the system) having an impact on its industrial performance (Naciri et al, 2012b).

Logistics environment: In addition to the reliability, efficiency and reactivity levers, the eco-logistics lever focuses on respect for the environment and societal development by limiting the pollution caused by logistics activities and promoting the development of territories (SC Meter, 2015).

2.3 Supply chain performance measures

Performance measurement is not a recent concern. The advent of the industrial age only amplified the need for performance measurement, particularly in assessing mass production levels, but it was undoubtedly in the 1980s, with the onset of the information age and new technologies, that performance measurement really took off in companies (O'Brien, 2000).

2.3.1 Logistics performance evaluation approaches

The evaluation of logistics performance is one of the major challenges that modern companies have to face, some of them being :

- customer service (including the improvement of service levels),
- strategic partnerships as a lever for logistics integration,
- inventory management,
- the management of physical and related information flows,
- cost control, cycle time reduction,
- Geographical coverage and flexibility.

These challenges emerge mainly because of the decentralisation of firms' production systems induced by their reorientation towards the development of their core competencies and by the need to implement efficient logistics integration mechanisms (Gelinas, 2002).

Therefore, the performance of a company should be assessed in relation to its role in the supply chain to which it belongs, but above all in relation to the objectives that allow it to position itself within that supply chain.

Undoubtedly, re-evaluation of logistic performance is a strong current trend that involves various measurements and estimations: establishing the level of performance according to several logistic indicators, establishing the links between the results from the indicators and the logistic objectives of the company and determining the contribution of the level of achievement of logistic objectives to the achievement of the strategic objectives of the company as elements of definition of its orientation towards competitiveness

In recent years, several approaches have been put forward to assess logistics performance. Table 2.1 presents some of the approaches that help companies to improve their logistics performance by summarising their main characteristics.

Table 2.1 Logistics performance evaluation approaches

Model	Features
ASLOG (Association frangaise pour la Logistique)	Scorecard questionnaire Internal but not external benchmarking Evaluates logistical procedures Analysis of the strengths and weaknesses of these procedures
SCOR (Supply Chain Operations Reference)	Evaluates key supply chain management processes Strategic and operational assessments External benchmarking against best practice Identifies areas for improvement Provides software mapping to achieve best practice
TBP (Balanced Scorecard)	Indicators to target performance improvement More of a strategic level Identifies the determinants of long-term performance improvement Evaluates both financial results and customers, internal processes and organisational learning
EVALOG (Logistics Evaluation)	Define "the basic requirements for assessing the quality of logistics performance"; The standard aims to enable its users "to identify the areas where they need to improve in order to manage their physical and logical

	flows". The originality of EVALOG compared to other guides is that it is "a single document common to suppliers and customers" of the European automotive industry; EVALOG can be used by "large, medium and small companies" either for "
	This can be done either as a "self-assessment of their logistics performance" or as an "audit document".

These standards, created or recognised by industry, have all evolved to include environmental and social considerations in addition to economic and financial ones.

2.3.2 Steps to implement a logistics performance measurement system (LPMS).

In the literature, we find authors who have proposed approaches to implement a logistics performance measurement system (LPMS). Among these works we can cite those of :

> **Laverty and Demeestere:** The authors outline the principles for developing a system of performance indicators that tracks costs and performance in a JIT logic: act on causes; provide a common, clear and explicit language; focus on controllable elements; ensure coherence of actions and their convergence towards strategic objectives; retain a limited number of priority indicators; adapt the monitoring period to the problem to be addressed; and create a dynamic for improving performance (Mauchand, 2007)

To select performance indicators, cause and effect analysis is used, linking an objective to one or more key variables. Then performance indicators are set that measure the achievement of the objectives and steering indicators that monitor the action variables (Laverty, Demeestere, 1990).

> **Bourne:** They propose an approach for the SMPL project consisting of three key phases (Bourne et al., 2000):

- The design of performance measures involves two steps: determining the objectives to be measured and defining the measures themselves.

- implementation of performance measures ;

- the use of performance measures which involves two steps: using the measures to implement the strategy and using the results of the measures to test the validity of the strategy.

Thus the authors place their approaches within the framework of using the SMPL as a tool for continuous improvement (Measure, Review, Act) which touches on objectives, measures and strategy.

> **Capar:** The author insists that the implementation phase of a performance measurement system is important, and that it must be conducted carefully. He proposes eight steps for implementing a supply chain performance measurement system (Capar, 2002):

- identifying strategic directions ;

- analyse the existing performance measurement system ;

- choosing the right indicators, determining the KPIs (Key Performance Indicators);

- choosing the appropriate measurement method ;

- achieve understanding and acceptance of the performance measurement system;

- set targets for all indicators;
- eliminate contradictory indicators.

> **Berrah and Vincent:** The authors propose five steps for implementing a performance measurement system (Berrah and Vincent, 2007):

- analyse the existing situation, translate the strategy into overall objectives and select the major processes for achieving these objectives;
- identifying critical activities and performance factors ;
- Analyse performance and choose action variables, Set up performance indicators;
- define the appropriate aggregations to measure the overall performance of the supply chain.

> **Okar:** In order to successfully implement a performance measurement system within a supply chain. The author proposes the following six phases:

- preliminary phase of the project ;
- preparing human resources ;
- choice of model ;
- definition of the SMPL ;
- launch of the SMPL ;
- development of the SMPL (Okar, 2011).

2.3.3 How to improve the performance of a supply chain?

Traditionally, the most common practice is to measure the financial aspect which is easy to do and look at the balance between revenues and expenses. The problem is that financial metrics are inadequate to measure the performance of a supply chain because they do not have a clear view on efficiency at the operational level (Camirenelli and Cantu, 2006), and do not take into account the customer's quality of service. It can even be said that decisions at the operational level are better managed if the financial aspect is not taken into account. But for a global optimisation of the supply chain, a balance has to be found between financial and non-financial metrics, whether at strategic, tactical or operational level. Also, to be able to make the right measurements in all functions of the chain would allow to better understand it and thus to be able to improve it where the needs are felt. After discussing the measurement of supply chain performance, the following section will elaborate on the concept of risk and risk management in the supply chain.

2.4 Risk and risk management in the supply chain

2.4.1 Definition

L'Houssaine (2013) considers the notion of risk in the supply chain as *"incidents/events that are unpredictable, affecting/originating from one or more partners of a supply chain and/or its processes, which can negatively influence the achievement of the organisations' objectives".*

Risk therefore has two facets: a danger to be avoided and an opportunity factor. The decision-maker examines the risks associated with the set of possible strategies to be deployed and chooses from this set to select a strategy while being aware of the risk it carries.

Supply Chain Risk Management (SCRM) therefore aims to manage supply chain risks. According to (Artebrant et al, 2003) SCRM is *"the identification and management of risks originating from within or outside the supply chain, through a coordinated approach, involving the members of the chain, and seeking to reduce the vulnerability of the chain, i.e. the supply chain, as a whole"*.

What emerges from this definition is the double view that one could have on the risks of the supply chain: an internal view - that is, what characterises the risks observed within the chain - and an external view - that is, what characterises the risks coming from the environment in which the supply chain evolves, in particular the risks related to the uncertainty of demand. Furthermore, this definition specifies a fundamental aspect of SCRM: the reduction of the chain's vulnerability, and emphasises the fact that SCRM is a collective action of the various actors in the chain and not an isolated action carried out by one actor in the chain (Jaouhar, 2006). This highlights the need to establish cooperative relationships between actors in order to conduct a successful risk management process.

2.4.2 Risks associated with supply chains and influence on performance

L'houssaine (2014) addresses the issue of risks associated with CL and their influence on CL performance. The empirical results, which they conducted among a large sample of Moroccan industrial firms, confirm that the risks in question negatively influence CL performance.

> **Upstream incidents**

The upstream failures come from the suppliers themselves (failure of logistic performance, problems of production incapacity due to important variations of the company's orders), and from the conditions on the supply markets (shortage of products, bottlenecks, fluctuations of exchange rates).

> **Internal incidents**

Operational risks have a significant impact on the efficiency of logistics flows, i.e. on the management of product flows within the company. These infrastructure-related risks arise from the use of the SC's logistics equipment. The unavailability (breakdown, stoppage, strike, computer virus, software error, etc.) of the infrastructure at the desired time can lead to lost opportunities and inefficiency. This can lead to delays, quality and quantity problems in production and stock-outs (Narasimhan and Talluri 2009; Tang and Nurmaya Musa 2010). They relate to potential infrastructure failures for technical, human or infrastructural reasons.

> **Downstream incidents**

Downstream of LC, failures arise from the unpredictability of the downstream market due to the volatility of demand, making sales forecasts difficult. This unpredictability can lead to problems such as disruptions in the transport and physical distribution of products to customers. These incidents are caused by problems of information asymmetry between the company and the customers. They also arise from elements associated with the business relationship such as late payment or cancellation of firm orders by customers (Van Der Vorst and Beulens, 2002).

2.5 The integration of risk management into the logistics performance measurement system (LPMS)

The definition of the SMPL represents an opportune moment to integrate the risk management component (Okar, 2011). Recent studies have focused on the interaction between performance measurement and risk management in a supply chain, and have recommended integrating both processes into supply chain management in order to improve performance and reduce or avoid risk (Horvath, 2003; Winkler and Kaluza, 2006; Ritchie and Brindley, 2007). According to Mikus (2001), the following steps should be taken:

■ Definition of a common target system, identification and quantification of the associated potential risks;

■ Planning supply chain strategies taking into account potential risks already identified;

■ Select appropriate measures for supply chain management and risk management;

■ Implement performance and risk measures and adapt them to each company;

■ After the measurements have been made, a detailed verification of the impact of the performance and risk situation using the performance and risk parameters already defined;

■ Integrated assessment of performance and risk parameters and decision on appropriate steering measures.

An SMPL should include indicators that measure and alert decision makers to disruptions in physical, financial and information flows. It should be an effective tool for steering the supply chain to prevent a range of risks from occurring and to help minimise the damage if they do occur.

For supplier risk management, the performance measurement system can provide a range of useful and relevant information: Importance of the supplier, Quality of the supplied items, Price competitiveness, Quality of the service, Payment conditions etc. In the same way, the performance measurement system helps to manage customer risk through several indicators:

■ Sales growth by region, by channel, by product type,

■ Evaluation of customer satisfaction, Follow-up of customer complaints, Number of new customers, Number of lost customers, Follow-up of customer credits etc.

2.6 From measurement to performance indicator

2.6.1 Definition

The characteristics of a performance indicator are reflected in the following definitions:

1. A performance indicator is a quantified data that expresses the effectiveness and/or efficiency of all or part of a system (real or simulated), in relation to a standard, a plan determined and accepted within the framework of a business strategy (Biteau et al., 1991), (Berrah, 1997).

2. A performance indicator is a numerical translation of the strategic objectives pursued by the organisation (Epstein and Manzoni, 1998).

3. A performance indicator is information that should help an individual or an organisation to steer the course of an action towards an objective, or that should enable him to evaluate the result. (Bonnefous, 2001).

4. A performance indicator is associated with an "action to be piloted" whose operational relevance

it must reveal (Lorino, 2001), (Bouquin, 2004).

The indicator is therefore seen as an "objective measure" (Bitton, 1990), a decision-making element that makes it possible either to control processes with a view to achieving defined objectives (control logic), or to modify the objectives themselves (progress logic). (Cohen and Russel, 2005) recommend a performance management approach based on the quality of the performance indicators chosen: each indicator must be closely linked to the company's strategy, comprehensible and relevant.

As illustrated in the figure (Figure 2.2), an indicator is derived from the target objective and the knowledge of the levers for action. It is therefore interesting, according to the authors, to properly identify the objectives to be reached by internal benchmarking (comparison of performance within the different units of the same company) and external benchmarking (positioning of results in relation to the industrial context and to competitors) in order to identify opportunities for improvement. A more detailed presentation of Benchmarking is given later.

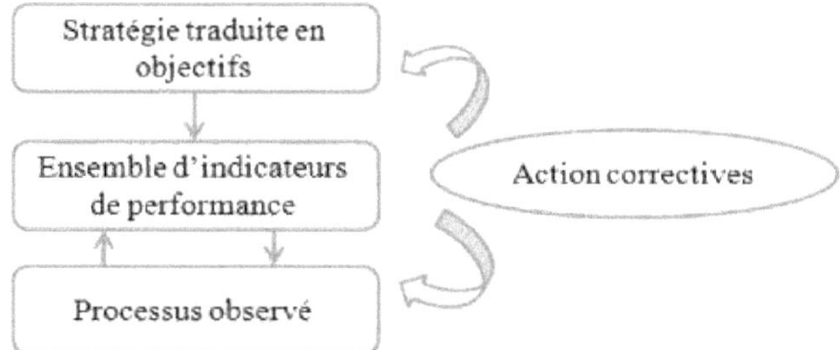

Figure 2.2 Implementation of performance indicators (adapted from Lorino, 2001)

2.6.2 The quality of performance indicators

According to Lorino (2001), the relevance and quality of performance indicators are assessed along three dimensions:

Firstly, **the strategic relevance** of the indicator: the indicator must be associated with a strategic objective to be achieved. It provides information on whether or not an action is being carried out correctly and is contributing to the achievement of the objectives. An indicator that is unsuited to the target objective may be counterproductive and may lead to derives.

Secondly, the quality of an indicator is based on **its cognitive capacity**. The indicator must be able to "signal", to easily guide the actor, or more generally the group of actors, to act and understand the factors of success or failure. When reading it, the decision-maker(s) must be able to act and be motivated to act.

Finally, the last criterion for assessing the quality of an indicator is **its operational relevance**. This consists of verifying that the measurements taken are the results of a specific and identified type of action, and that the data used are trustworthy. The operational relevance of an indicator therefore concerns the validity of the results. The relationship between indicator and action must be unidirectional: from action to indicator. The indicator is deduced from the choice of action (the indicator is only useful for steering the action and its result) and not the other way round.

Performance indicators therefore interact with three components: the objectives induced by the strategy, the actors who are the recipients of the information, and the actions taken by the actors to achieve the objectives (Figure 2.3).

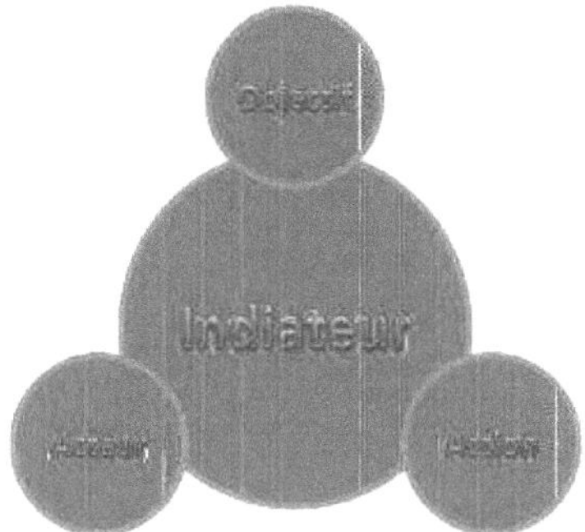

Figure 2.3 The indicator 'triangle': strategy translated into objective, action process and collective actor (Lorino, 2001)

Fernandez (2005) completes this list of characteristics with three other criteria:

> A performance indicator should be easy to construct, with no major difficulties impeding its realisation. The construction of a performance indicator should be carried out at an acceptable cost.

> The cost of development and construction must always be compared to the decisional value of the door mess.

> The information provided should be updated on a cycle that is specific and tailored to each performance indicator to allow for truly effective decision-making.

2.6.3 Nature and typology of indicators

Indicators can be classified according to two main types: quantitative indicators and qualitative indicators. Quantitative indicators are used to quantify a measure, i.e. a measure of a physical quantity. In this type of measurement we find quantities such as: distribution cost, production cost: capacity of machines, energy used, etc... They can be expressed in relative form.

To relativise and express quantitative measures, several numerical expressions can be used:

■ **Ratio**: It allows to express relationships of order of magnitude between elements which may or may not be of the same nature.

■ **Rate**: A rate is a ratio that combines measures of different natures.

■ **Index**: The index is a ratio, as it is also a dimensionless number that expresses a ratio between two elements.

Qualitative indicators are an assessment on a subjective value scale, such as: customer satisfaction, customer response time, machine flexibility, labour flexibility (Moutaouakil, 2015)

The relationship between qualitative and quantitative indicators can be mixed since qualitative measures can be used to construct quantitative indicators and vice versa.

There is a lot of literature on the different types of indicators. Depending on the field of management, the names of the different types of indicators may vary. There is no real consensus on the terminologies used. However, there are many names that can have the same meaning. Two main categories of indicators seem to emerge, **advance indicators** and **outcome indicators** (Juglaret, 2012).

Outcome indicators are so-called "reactive" indicators, after the event. They are sometimes also called 'effect', 'impact', 'efficiency' or 'impact' indicators. Leading indicators are defined by Wreathall (2009) as indicators that inform about a change before the economy itself has changed.

2.6.4 Measuring supply chain performance through a system of indicators Three main supply chain performance indicators are widely used, each corresponding to a type of flow: "cooperation" indicators for the performance of the information flow, costs for the financial flow and delivery times for the physical flow. The first step in performance monitoring is therefore to "measure performance".

Several performance criteria can be considered. Beamon (Beamon, 1998) classifies these into two categories:

- qualitative performance measures (customer satisfaction, flexibility, integration of physical and information flow, financial risk management, etc.);

- quantitative performance measures (late delivery, customer response time, etc.).

Secondly, reengineering decisions need to be made and the system and model need to be influenced by decision variables in order to achieve the set objectives, as shown in Figure 2.4. The implementation of an efficient system therefore reflects a need to control the supply chain and improve performance.

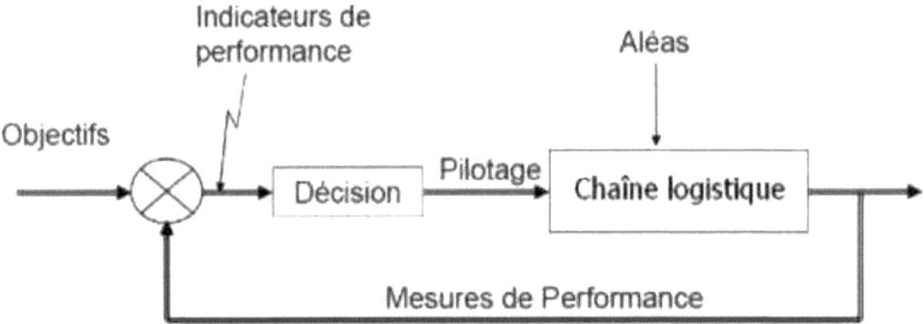

Figure 2.4 Control system for LC performance

2.6.5 Methods of defining and implementing a system of performance indicators

This section provides an overview of the most common methods of defining and implementing performance indicators. The objective is to understand the different views of analysis and their implementation for performance evaluation. We will present the best known methods, which have been proven in industrial applications (see ABM-ABC, SCOR, ECOGRAI, BSC and Benchmarking) presented in chronological order.

2.6.5.1 ECOGRAI (1990)

The ECOGRAI method, developed by (Bitton, 1990), makes it possible to design and implement systems of performance indicators for the purpose of re-evaluating the technical and economic performance of the enterprise's production system or one of its parts (Figure 2.5). The steering structure refers to the GRAI grid (Doumeingts, 1984) which shows the decision centres for each level (strategic, tactical and operational) and each function of the production system. It is based on the triplet objective, measure, variable in order to design and implement in all decision centres a system of performance indicators. It is a participatory approach that involves future users in the definition of indicators at all levels of the hierarchy. It aims to achieve a limited number of indicators per function and per level.

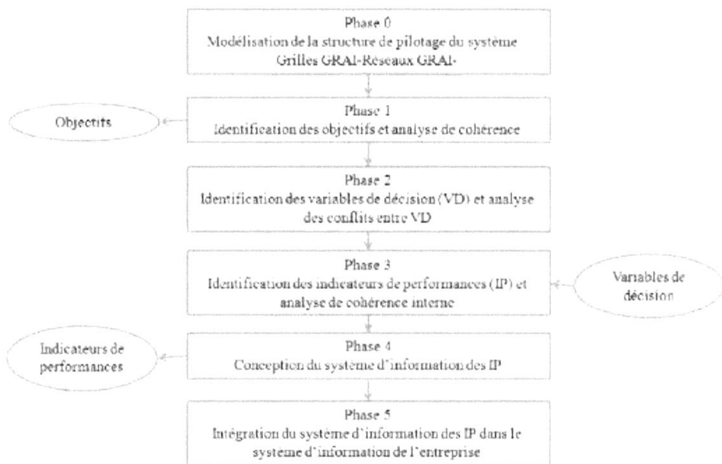

Figure 2.5 The six phases of the ECOGRAI method (Bitton, 1990)

The objectives are broken down by level and by function. Each objective is defined in coherence with the decision variables in order to ensure control over the production system on which it acts. Measurements are made after the decision has been made by performance indicators in order to verify the results of the decision. It should be noted that the coherence between decision variables and objectives is very important to take into account, and has been described and developed in the work of (Ducq, 2003).

The ECOGRAI method has been used by several researchers to facilitate the determination of performance indicators. We cite here the work of Frederic Bonvoisin (2011) who used the ECOGRAI method to develop performance evaluation tools for hospital operating theatres. In Mouss et al (2004), the authors used the method to develop performance indicators to improve the classical management approach of a production system for the company Laiterie Aures. In Robin et al (2005), the authors proposed a performance evaluation model to evaluate a product design system and monitor its evolution. In Bitton (1990), the author used the ECOGRAI method for the design of a dashboard structure supporting a factory with a high degree of automation.

The originality of the ECOGRAI method lies not in the definition of performance indicators, but in the approach: 1 Objectives, 2 Decision variables, 3 Performance indicators (Verane, 2008).

2.6.5.2 The më-thode ABC-ABM: activity-based management (1991)

Activity Based Costing (ABC) and Activity Based Management (ABM) are approaches designed to provide relevant information on costs and margins. In particular, they make it possible to improve the use of available resources by shedding light on the choice of subcontracting, by helping to define the organisation of competences or by providing the company with performance-oriented dashboards (Ravignon et al, 1998).

The principle of activity-based management is ultimately to obtain the real cost of a product or service and, by extension, the real cost of a service:

- the cost of the components of the product or of each stage of its process;
- comprehensive and detailed budgetary control ;
- identifying dysfunctions between activities ;
- monitoring of deviations, overruns, by activity and by product ;
- costing simulation for the launch of any new product;
- identifying the steps to be taken to reach a profitability target.

This approach has many advantages. All analyses are conducted on the sole basis of costs, which are all treated as variable costs. Indeed, the costs of the products are followed through their consumption of activities, which integrate all the expenses in a direct and variable form. Indirect and direct costs no longer exist because all expenses are allocated to activities. Activities consume all expenses, and products (or services) consume all activities.

This approach requires a global vision of the company (activities that make up the processes), as well as the mastery of certain skills (management, project management).

The method consists of five steps (Opartners, 2014):

- the first phase consists of establishing a process map. The aim here is to identify the activities (which make up the processes) and the different outputs of the system;
- the second phase allocates workloads and time to the different activities. This constitutes the resources for the operation of the activity;
- the third phase aims to define, for each activity, a system of performance indicators measuring the output of the activity, generating costs;
- the fourth phase seeks to identify the quantity consumed by each product. The output of the activity is effectively associated with the consumption

Chapter 2: Measuring performance in the resulting resource supply chain. This gives the amount of resources consumed per product and the associated costs;

■ the last phase determines the product cost, in total cost and in unit cost, detailed by activities. This is a precise cost estimate that will allow the implementation of a budgetary control, the simulation of costs for new products, the analysis of variances and activities, the detailed analysis of components.

Activity-based management is based on the following three dashboards:

■ the activity dashboard (monthly) which allows the monitoring of the achievement targets;

■ the financial dashboard (monthly/quarterly) to monitor financial targets;

■ the structure dashboard (quarterly/semi-annually) which allows the monitoring of cost structures.

ABC and ABM emphasise the need to implement steering according to the processes (or rather, the activities that make up these processes) that define the system under study.

2.6.5.3 Balanced Scorecards (BSC)

Some traditional approaches to performance measurement ignore a dimension deemed crucial by (Kaplan and Norton, 1992), namely the consideration of the interactions between strategic objectives and operational performance, combined with the deployment of these objectives and performance at all levels of the organisation. Realising that no single measure can provide relevant performance, they propose the concept of 'Balanced Scorecards', which are based on a rigorous framework for expressing strategic objectives and a methodology for operationalising them.

The performance indicators are classified along four axes (Figure 2.6):

- The "financial performance" axis includes indicators such as product prices or supply costs, salaries, transport costs, value added productivity, capital turnover. Indeed, as noted earlier, financial indicators alone are relatively easy to measure but do not provide a complete picture of the smooth running of supply chain activities.

- The "internal process" axis contains indicators such as sales forecasts, production quality, production flexibility, internal cycle times. These indicators assess operational performance and are not necessarily linked to financial results.

- The "customer" axis contains indicators that determine customer-oriented performance such as on-time delivery, order execution cycle, customer satisfaction rate and order execution compliance.

- The "organisational learning" axis is the most difficult dimension to define; its indicators quantify the effectiveness of the company in integrating new skills.

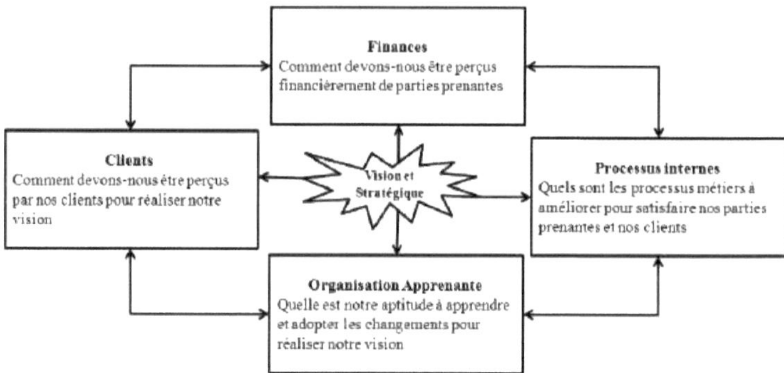

Figure 2.6 The four axes of the Balanced Scorecard (Kaplan and Norton, 1992)

2.6.5.4 Benchmarking (1999)

Another way of approaching the problem of performance evaluation is through benchmarking (Anderson et al, 1999), which describes benchmarking, or 'learning from others', as the following sequence of steps:

- Measuring one's own performance and that of benchmark organisations, with the aim of comparing and improving

- Comparison of performance levels, processes, and practices

- Learning from each other's good practices to introduce improvements in their own organisations.

- The implementation of performance-enhancing solutions is the ultimate goal of benchmarking.

Both internal and external benchmarking can provide valuable data for performance improvement. Intra-company benchmarking can identify which unit within the company is performing in an exemplary way.

Once internal indicators are generated and relevant benchmarking data are collected, inter-firm benchmarking is used to contextualise the firm in its industrial environment, in order to identify opportunities for improvement (Simatupang and Sridharan, 2004), (Wong and Wong, 2008). Typically, companies use external benchmarking to study competitors' industrial practices with the objective of improving their own performance (Simatupang and Sridharan, 2004).

It is possible to limit the comparison to industries of the same nature (similar characteristics in terms of product complexity, geographical distribution, production strategy, ...) (Gilmour, 1999) or to extend the analysis to industries with other characteristics. (Handfield and Straight, 2004) argue that both internal and external benchmarking efforts provide decision-makers with an interesting base of indicators to understand how to direct their efforts.

2.6.5.5 SCOR model (2000)

The SCOR approach also provides a set of performance indicators for each level of decomposition. Level 1 indicators are relevant for monitoring performance at the most aggregated level, but are of lesser use in diagnosing the causes of performance degradation. The more detailed performance measures provide precision to the analysis of dysfunctions. Thus, in line with the hierarchy of the models in SCOR, each level 1 indicator is associated with a group of level 2 and 3 indicators, which are useful for diagnosing the causes of dysfunctions identified at level 1.

2.6.5.6 **Literature review:**

The methods for implementing performance indicator systems (PIS) have grown in recent years to offer a wide range of tools for decision-makers. Each of these methods has its own particularities and implementation approach.

Following our study and literature review of the different methods of designing performance indicator systems, we observe that there are points of convergence between these methods. Indeed, several methods :

■ Consider the modelling, the clear expression of the system indispensable to the creation of a SIP, because one cannot pilot or evaluate a system whose structure is unknown (SCOR, ECOGRAI). *In this sense, we recommend the SCOR model as a tool for modelling the supply chain*

■ Agree on the need to set strategic objectives (e.g. TBP). However, they do not specify how strategic objectives should be established. *In this sense, the Balanced Scorecard method is of interest, as it identifies performance along four strategic axes (customer, internal process, organisational learning and financial).*

■ Insist on the deployment of objectives on all horizontal (processes) and vertical (from strategic to operational) levels (TBP, ECOGRAI, SCOR). *In this sense, we retain the deployment of the process approach, which consists of deploying the general objectives in a horizontal and vertical way on all the processes of the supply chain.*

As regards the analysis of the coherence of the indicators, we distinguish two main categories:

■ Those that set a pre-established framework of indicators (TBP and SCOR...), in which case compatibility is guaranteed by the logic of the approach;

■ Those who opt for the implementation of specific indicators for each company (ECOGRAI, ABC/ABM...), in this case it is essential to analyse and ensure the compatibility of indicators.

In addition, most methods do not propose a methodological framework for selecting appropriate indicators. *Here, we carry out a reflection on the choice of indicators combining those proposed by the SCOR model and those of our own referential, which is composed of logistical indicators considered important for companies.*

2.6.6 Logistics indicators

In the literature, we find references that focus on the evaluation of supply chain management via Key Performance Indicators (KPIs), such as

■ the work of Gunasekaran et al, (2001) where a list of 42 indicators is proposed, classified into strategic, tactical and operational. This work forms the basis for practical reflections (Morana, 2002; Bhagwat and Sharma, 2007; Sharma and Bhagwat, 2007);

■ the work of Gunasekaran and Kobu (2007) with a list of 26 indicators;

■ and the work of Griffis et alii (2007) with 14 indicators.

Based on the work of (Gunasekaran et al, 2004; SCC, 2011; Gunasekaran et al, 2001; Barut et al, 2002; Sahin and Robinson, 2005; Wuet Song, 2005; De Toni and Nassimbeni, 2001). France-Anne (2007) proposes a synthesis of indicators often cited in the literature to measure the performance of a supply chain. He distinguishes between strategic, tactical and operational level indicators, and links them to a macro-process, thus differentiating between internal and external indicators. At the strategic level, the indicators cover the different processes of the supply chain and are mostly transversal. On the other hand, at the tactical and operational level, each indicator is associated with a process.

- the KPI benchmark (KPI, 2015).

2.6.7 *Performance* indicator *standards*

The main benefit of standardisation is to give a common definition for each indicator. This has several advantages: it ensures that the indicator is well calculated, using a common calculation method and it facilitates benchmarking, allowing comparison of homogeneous measures.

Standards have been defined for performance measurement. They provide a comprehensive set of indicators, which are recognised as effective in companies.

In the SCOR-model, each sub-process of levels 2 and 3 is associated with a set of indicators according to 5 performance criteria: reliability (measured, among other things, by the percentage of goods on the shelf and the accuracy of replenishment), reactivity (measured by the replenishment lead time), flexibility, cost, stock level (number of days of stock in the warehouse) The SCOR-model introduces about 200 performance indicators for supply chain processes.

The performance indicators in the CEN Logistics (European Community of Standardisation) Logistics report focus on the description and codification of best practices in logistics. The performance indicators proposed by CEN are classified according to the different activities of the logistics chain, which are: sales and customer service, procurement and supplier service, product, production, storage, transport, inventory control, and various measures. About 100 performance indicators are defined and positioned in the different categories.

2.7. *Review of work on supply chain performance measurement and positioning*

The field of performance and its measurement in the supply chain has been the subject of several contributions. In the table (Table 2.2), we present a summary of recent work in this area and its industrial context.

Table 2.2 Recent work on measuring LC performance

Authors	Contributions	Context
Lohman et al, 2004	They propose indicator systems derived from the (Kaplan and Norton, 1992) approach based on a four-step structure.	French
Fynes et al, 2004	It proposes seven areas of performance to be considered: customers, partners, employees, sustainability, shareholders, internal processes and information systems. It proposes an extension of the approach of (Kaplan and Norton, 1992).	Franfais
Folan and Browne, 2005	This paper proposes a step-by-step process from recommendations for system analysis to the definition of a formal framework for the	French

	implementation of a performance system for effective supply chain management.	
Bhagwat, 2007	Proposes lists of useful indicators for assessing supply chain performance. Explains the four perspectives of the Kaplan and Norton TBP.	Franfais
Antony Valla, 2008	Proposes a general method for diagnosing the supply chain.	Franfais
Emilie Baumann, 2011	Proposes models for assessing economic, environmental and social performance in supply chains	Franfais
Chafik OKAR, 2011	Proposal of a maturity model for the development and implementation of a performance measurement system in a supply chain	**Moroccan**
Claude FIORE and Imane EL KARTIT, 2014	Logistics performance management using Lean management	Franfais
Ouabouch Lhoussaine, Lavastre Olivier, 2014	Vulnerability, risk and performance in Supply Chain Management - Case of the agro-food industry in Morocco	**Moroccan**

From this bibliographic synthesis, we notice a paucity of scientific contributions in the field of supply chain performance measurement in the Moroccan industrial context.

Indeed, (Okar, 2011) proposed a six-phase Logistics Performance Measurement System (LPMS) project management framework. This was based on a literature review. His proposal was validated in a large industrial company.

Since our problem is the definition of a dashboard as a model of the SMPL, we can note, at this level, some limitations in the contribution:

(1) Definition of the supply chain by department (by function) and not by process.

(2) The selection of indicators is not justified and does not meet the company's objectives. Indeed, it is the monitoring indicators that have been used and not the performance indicators.

(3) The SMPL is not used as a strategic alignment tool;

Other existing work focuses on empirical studies, such as management control, which allow the diagnosis of CL in large companies.

The overview and analysis of the literature led us to combine three methods of implementing indicators to measure the performance of a supply chain:

> The Balanced Scorecard philosophy is concerned with a holistic approach to performance, resulting in a balanced measurement system. The latter is balanced between financial and non-financial indicators, between short term and long term and between intermediate indicators and outcome measures. In this sense, Bhagwat (2007) suggests the assignment of a set of indicators informing about the performance of the supply chain to the TBP axes.

But the choice of indicators not appropriate to the situation of the company or to the results to be achieved could constitute a danger to the real stakes of the structure and the probability of complying with its strategy.

However, this approach does not provide a methodological guide for identifying the relevant indicators that feed into the supply chain scorecard. To remedy this shortcoming, we use two references for the definition of our indicators:

> The SCOR model

The Supply Chain Council's SCOR is a good reference because the methodology used analyses performance across five key supply chain processes: planning, sourcing, production, delivery, rework. It hierarchises the indicators according to the 3 levels: process, task, activity.

This model associates to each sub-process of levels 2 and 3 a set of indicators classified in 5 categories: reliability, responsiveness, flexibility, cost and resource management. And it introduces about 200 performance indicators for the supply chain processes.

> Our own bank of indicators

Our bank of indicators is based on a synthesis of the literature and a selection through a questionnaire survey. This bank includes only those indicators that are relevant and useful for Moroccan industries.

2.8 Conclusion

The mission of supply chain management is to satisfy customers by meeting commitments in terms of quality, cost and delivery time. Although the importance of supply chain performance within the performance of companies is growing, its measurement is no less delicate.

In this chapter, we first :

- Addresses the concept of performance and performance measurement;

- Situates the notion of performance measurement in the context of the supply chain; Here, we have proposed to define the notion of supply chain performance by the degree of achievement of the objectives set by the supply chain management missions.

Then we have :

- Situates the notion of risk and its management in the emerging field of supply chain management.

- discusses the concept, quality, and typology of performance indicators, while citing its main implementation methods.

At the end, a review of the literature on supply chain performance measurement was presented, as well as our position on the recommended methods for implementing our performance indicators.

The next chapter will be devoted to the introduction of the concept of the dashboard, the main approaches and the practical methodologies of its construction.

Part 2

Supply chain management by dashboard and methodology proposed

In order to evaluate the progress of the company, it is considered essential to have a tool for steering its supply chain, in this case the dashboard. The latter is a measuring instrument that allows the performance of a system to be monitored and controlled.

In this part, we will discuss the management of the supply chain by means of a dashboard and we will present our method of management of the supply chain adapted to the industrial context.

Chapter 3: Supply Chain Management by Dashboard

3.1 Introduction

In this chapter, we start by describing what a dashboard is and introduce the main approaches to designing a dashboard: the OVAR approach and the Balanced Scorecard (BSC) approach. Then, we will detail the practical methodologies for implementing a dashboard.

Given the importance of having a balanced scorecard in industrial companies, we will conduct a literature review on its use in the supply chain.

Finally, a synthesis of recent studies on dashboard-based management will be presented and discussed in order to position our research work in relation to what already exists.

3.2 The concept of dashboards

3.2.1 Definitions

The concept of the dashboard appeared in France at the beginning of the 20th century[eme] . It was mainly developed by engineers with very technical functions. The latter were looking for methods to improve production processes and to better understand the cause-effect relationships between the actions implemented and their impact on the performance level of the various processes (Juglaret, 2012).

In the literature, the dashboard is presented as a decision-making tool (Malo, 1992), or as a management instrument that conforms to a pyramid-type organisation (Guerny et al., 1990), structured on the logic of objectives-key variables-indicators (OVAR method), and oriented towards the remote steering of delegated responsibilities (Malo, 1992).

Malo (2000) defines the dashboard as 'a tool for the top management of a company which allows a global and synthetic view on the state of the operations in progress and on its environment'. Until the 1980s, the dashboard was seen as a "reporting" device, allowing the level of achievement of previously set objectives to be monitored (Ardoin et al., 1986), but the 1990s saw the dashboard evolve towards a more action plan oriented approach which led to the OVAR (Objectives, Action Variables, Accountability) method.

Bouquin (2001) defines the dashboard as "a set of few indicators (five to ten) designed to enable managers to take cognisance of the state of revolution of the systems they control and to identify the trends that will influence them over a period consistent with the nature of their functions".

Recently Groger (2013) distinguishes between two meanings of dashboard: the narrower sense where the term refers to tools for graphical visualisation of key performance indicators and the broader sense where the dashboard is used for monitoring, analysing and optimising critical business activities enabling users at all levels of the hierarchy to improve their decisions.

3.2.2 Functions and characteristics of a dashboard:

A good dashboard is not only worth the sum of the indicators it represents.

According to Daum (2005), the dashboard should contain information on

- The success factors that contribute to the achievement of the defined objectives (the action variables).

\- The measures and initiatives implemented to achieve the objectives or milestones (action plans).

\- Performance levels at a granular level for each unit of an organisation with regard to the achievement of results.

According to Berland (2009). It should enable :

\- Getting people to talk about the strategy so that it can be implemented better.

\- Provide everyone with a common performance monitoring tool (reporting and self-monitoring).

\- Understanding performance in its different aspects. The TBP has particularly deepened this dimension of the scoreboard.

According to Fernandez (2007), in contrast to a report, the dashboard does not only display the latest results, but it is first and foremost the management tool to fulfil several important functions in this field. These functions are therefore mainly related to management aspects rather than to simple tool functions helping management. Thus, according to (Juglaret, 2012) the dashboard allows to Reduce uncertainty, stabilise information, facilitate communication, stimulate reflection and control risk.

3.2.3 Choice of indicators :

The dashboard is a set of indicators but not just any indicators. According to Berland (2009) its indicators must fulfil the following conditions:

• They have to measure states and developments.

• They are used to control a system, i.e. a set of interacting elements within clearly defined boundaries.

• They must act within a time frame related to the pilot's obligations.

The dashboard is made up of indicators chosen by the decision-maker. These indicators remain limited in number, a maximum of ten, and allow an assessment of a situation. The dashboard also helps to understand and guide decision-makers in the implementation or not of corrective actions.

3.3 Dashboard Building Approaches (TDB)

The literature on performance measurement and management refers to two classic approaches to designing dashboards: the OVAR (Objectives, Action Variables, Responsibilities) method developed by the HEC Group and the Balanced Scorecard (BSC) by Kaplan and Norton (1998). We will therefore describe these two approaches in detail.

3.3.1. The French approach: The French Scoreboard (OVAR)

The OVAR method (Objectives, Action Variables, Responsibles) aims to deploy the strategy within the organisation by building the link between strategic objectives and action plans at the different levels of the organisation's hierarchy. It is thus in line with the classic principle of strategic alignment of the management instrument in the field of management control.

3.3.1.1 Concept and operating principle

The OVAR method is based on three concepts: objectives, action variables and responsibilities.

■ The objectives

Feisthammel and Massot (2005) define objectives as "very precise states to be achieved, for an identified performance criterion". The objective expresses the intention, the commitment, what one wants to achieve and produce. As evoked by Loning and Pesqueux (1998), "the objectives are the quantified and dated, operational declination of the general goals or missions incumbent on the managers. Their formation must be clear, precise, situated in time and must be able to be measured or at least evaluated "objectively"". The objective is stated explicitly (quantitative results or qualitative results) and must not be confused with the actions that lead to its achievement.

■ Action variables

An action variable is a variable that is causally related to the objective being sought. Action variables should not be confused with objectives. They should not be numerous so that the manager can concentrate his or her actions on the variables that are really causally linked to the objective sought. It is therefore essential to focus on those policy variables that have the greatest effect.

■ Responsibilities

The definition of objectives and action variables per responsibility centre depends on the organisational structure of the company. The objectives and action variables should not be outside the manager's scope of action in order for the manager to be able to act. As Mendoza et al (2002) point out, 'the implementation of scorecards risks being rejected if they are not consistent with the structures in place'. The OVAR method thus implies a reflection on the objectives pursued and the processes or activities on which efforts are concentrated at each level of responsibility in the company.

The elaboration of a dashboard according to the OVAR method is done in several steps:

■ The definition of the mission and objectives of the organisational unit through the organisational strategy;

■ Identifying the action variables available to responsibility centre managers;

■ Analysis of delegation and allocation of responsibilities;

■ The identification and choice of relevant indicators relating to both objectives and action variables in a logic of strategic alignment.

In the OVAR method, the scorecard is a tool for implementing the organisation's strategy. A causal model links the use of resources and the performance of the company. The strategy is translated into objectives and action variables by responsibility centre. The process starts with the formulation of objectives and action variables for the whole company. Subsequently, each management unit manager (responsibility centre) participates in the development of his or her scorecard. The delegation of objectives and action variables is therefore carried out following a debate between the managers in order to decide on the fields of responsibility of each. In practice, the method is based on coherence grids (objectives, action variables, responsibility centre) and where the action variables of hierarchical level N become objectives for hierarchical level N-1 (Bourguignon et al. (2001). The method thus makes it possible to declare the general objectives in a coherent manner in accordance with the organisational structure of the company. The choice of indicators depends on the objectives to be achieved and the action variables intended to achieve them (Distler, 2009) (Figure 3.1).

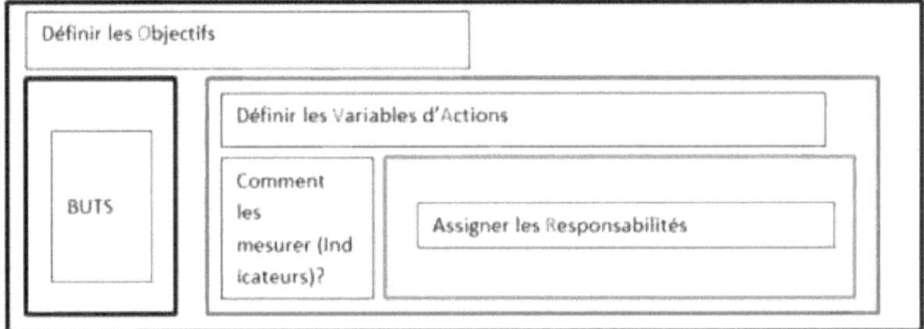

Figure 3.1 The fundamentals of the OVAR method

3.3.1.2 limitations of the OVAR method

We can point out some limitations in the design of an OVAR TDB such as :

> The risk of incompleteness in terms of action plans;

> An increased need for dialogue and negotiation;

> No guidelines to ensure the quality of the indicators.

However, we can say that it is the use of the tool and the way it is deployed and the degree of ownership that will determine the effectiveness of the results.

3.3.2 The American approach: the "balanced scorecard

The concept of the 'balanced scorecard' was described in 1992 by two American engineers: Robert S. Kaplan and David Norton. This section is organised into two sub-sections. First, the methodological framework that allows the construction of a balanced scorecard and the operating principle are explained. Finally, the identified limitations and shortcomings of the 'balanced scorecard' concept are discussed.

3.3.2.1 Concept and operating principle

For a long time, company performance has been limited to the assessment of financial performance. Several studies have shown that in practice, dashboards tend to be too often oriented on financial measures (Epstein, Manzoni, 1998). Financial indicators alone are not sufficient to guide a company's strategy in a competitive environment, as these indicators that measure past performance do not reflect the value created or lost during the last accounting period. Financial indicators do not provide enough information on the actions to be taken or those that have been taken to create financial value in the future. Based on this observation, Norton and Kaplan (1992) proposed a multi-criteria approach to performance organised along different strategic axes allowing better anticipation and control of the company's results.

The implementation of the Balanced Scorecard is based on a structured approach of four steps:

> Clarify and translate the project and strategy;

> Communicating and articulating ;

> Plan and define quantitative targets;

> Feedback and strategic follow-up.

43

The 'balanced scorecard' tool allows a strategy to be implemented along several dimensions. In general, Norton and Kaplan (1996) recommend the use of four general axes (Figure 3.2). These different dimensions allow the grouping of key elements and processes of the company for value creation. Consequently, each of these axes groups several indicators. These indicators represent the state or evolution of the level of functioning of the organisation's activities.

Figure 3.2 The strategy is structured along four general lines (Kaplan, 1996)

The different axes considered by default are therefore :

> *The financial axis*: this allows for an effective evaluation of the quantifiable economic effects of past actions. This axis determines whether the intentions and implementation of the strategy are proving effective (example of financial indicators: profitability, turnover, profits, etc.).

> *The customer axis*: this axis allows for a segmentation of the targeted markets. The indicators and determinants related to this dimension reflect the future financial performance of the company (examples of indicators: customer satisfaction, profitability by segment, loyalty, etc.).

> *The internal processes axis*: this axis leads the company to identify the key processes in which it must excel; the existing ones and those to be integrated. These are the processes that attract and retain customers in the targeted market segments and ensure the financial returns expected by shareholders. The processes can be innovation (design and development of a new product) or production (manufacturing, marketing, after sales service, etc.). In order to evaluate the proper functioning of each of these processes, several indicators will be created: production time and quality, delivery time, etc.

> *Organisational learning and development*: This is about the infrastructure that the company needs to put in place to improve performance and generate long-term growth. Organisational learning has three components: people, systems and procedures. It identifies the gap between current capabilities and those needed for true performance improvement. It helps to create a climate for change, innovation and development. The indicators used can be partly the same as for the customer axis (satisfaction, loyalty, etc.).

For the implementation of the Balanced Scorecard, the strategy is constructed from "strategy maps" (Figure 3.3). These maps contain the different axes for assessing the performance of the organisation.

44

On each of these axes, the determinants of performance are placed. The causal relationships in the achievement or improvement of these drivers are determined through hypotheses. Each indicator selected should constitute an element in the chain of causal relationships.

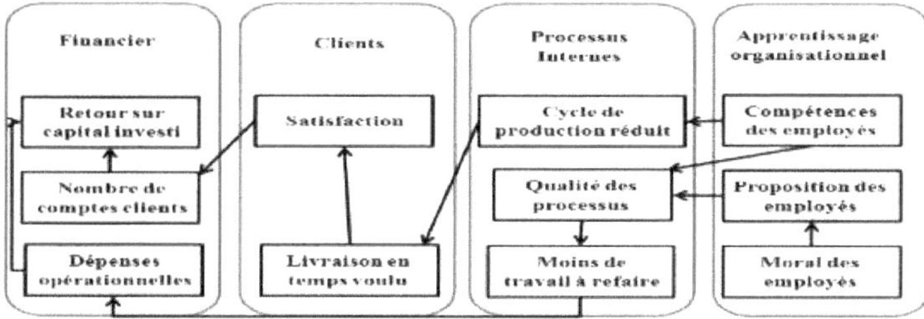

Figure 3.3 Example of a strategy map

In short, a 'balanced scorecard' helps to define the strategy, identify the drivers of performance and the relationships between the different strategic axes, as well as the new processes to be considered. It is much more than a simple performance measurement tool. It is also a vehicle for communication, mobilisation and awareness of the company's strategy throughout the organisation.

3.3.2.2. Limitations and shortcomings of the 'Balanced Scorecard

The shortcomings of the TBP include

o lack of clarity on the differences between the political and strategic dimensions. The achievement of a mission or vision has a strong political connotation, whereas the achievement of a goal implies a consideration of very strong economic criteria.

o The "customer" and "financial" axes most often group together the objectives to be achieved, while the other two axes generally group together the means to achieve these objectives.

o The cause and effect relationships between each of the axes are not necessarily obvious. Norton and Kaplan seem to overlook the interdependencies that may arise between each of the axes (Norreklit, 2000). The causal relationships between the determinants of performance on each axis are not necessarily unidirectional.

These limitations should be seen as areas for improvement to be taken into account in the further conceptualisation and development of our own dashboard.

3.3.3 Synthesis:

We summarise this paragraph by highlighting the main similarities and differences between the two approaches to building a TDB (OVAR and TBP).

There are many similarities between the OVAR method and Norton and Kaplan's TBP. Both methods: aim to better manage the implementation of strategic choices by ensuring the achievement of objectives with the help of relevant indicators;

are intended to avoid purely financial performance measures; encourage anticipation rather than reaction;

insist on selective measures to avoid information overload. apply a top-down approach in their construction processes.

However, the causal links are also found to be more 'open' in the franchise method than in Norton and Kaplan's TBP. The latter assumes causality between the four domains of the measures, and the existence of a general model of performance, which are relevant for all types of companies (Norreklit, 2000). The American approach does not assume any systematic or external causality:

In terms of the differences between the two methods, we find that :

o In TBP, the leader builds a roadmap for the set of measurement objectives and it is up to the whole organisation to align with the overall strategy and vision, which means that his or her subjectivity and the environment play a major role in designing the management system. In contrast, the OVAR-based dashboard is built up as it moves from a hierarchical level to a lower level.

o Kaplan and Norton's TBP approach is more helpful in promoting and disseminating the multidimensional view of performance, while the OVAR method emphasises the notion of intra-organisational communication of dashboards with the creation of multiple dashboards.

o The TBP is a ready-made device compared to the French approach, which requires the manager to adapt a very general process that requires managers to review the four axes by providing a framework for study.

Since our reflection is concerned with the development of a global supply chain dashboard performance management methodology, we believe that the TBP offers more inspiration and reflection. Thus, the OVAR method has a crucial advantage, which is the adhesion of all participants in the design of a dashboard. Indeed, the differences between the two approaches are the way in which both methods closely integrate objectives, strategies and performance measures within a causal framework and the way in which they are deployed.

3.4 Presentation of TDB design methodologies and practical guides

In addition to the performance measurement approaches presented above, other authors (Selmer, 2003; Fernandez, 1999; Voyer, 2000; Boix and Feminier, 2003), in the literature, have preferred to propose more practical methodologies organised in phases and steps, to guide the designer in his or her dashboard building process.

3.4.1 The Fernandez (1999) approach (GIMSI)

The GIMSI approach introduced by Femandez (1999) is intended to be a guide for the design of an SMP. The basic principles of this approach consist, on the one hand, in engaging in a reflection on the strategy and objectives of the company before moving on to the system design phase. On the other hand, the author also insists on the choice of indicators which must be in line with the local and global context of the organisation.

In other words, avoid using general indicators that do not apply to the specific context of the company. This approach is structured in four main phases divided into 10 steps, which are briefly described below (Table 3.1):

Table 3.1 Main phases of the GIMSI process

Phases	Description
Identification	Is carried out in two stages which consist firstly of identifying the company's

	environment in terms of the market and strategic[1] . secondly of studying the internal structure at the level of processes and professions.
Design	This is the core of the GIMSI process. It is carried out in four stages:
	1) defining local objectives that are aligned with the company's overall strategy,
	2) the study of the scoreboard as a whole,
	3) the choice of indicators,
	4) data collection
	5) the construction of the system.
Implementation	Is much more interested in the technical aspect of choosing the most suitable software package as well as in the integration and deployment of the decision support system with the other information systems of the company (customer relationship management (CRM), enterprise resource planning (ERP), etc.);
Audit	The aim of this phase is to keep the system in a permanent state of flux so that it always reflects the reality of the company's context.

3.4.2 Dimarche de Voyer (2002)

In his book, Voyer (2002) proposes an approach for the creation of a scoreboard, which includes five stages grouped into three main phases, more or less sequential: organisation and management of the project (stage 1), development or adaptation of the scoreboard (stages 2, 3, 4) and implementation and use of the scoreboard (stage 5) (Table 3.2):

Table 3.2 The main phases of the voyer approach (2002)

Steps	Description
Stage 1	the organisation of the project consists of studying the relevance of using a scoreboard and the opportunities it may present; identifying the target sectors and then studying the feasibility of committing resources in order to determine the scope of the project.
Step 2	the identification of management concerns and indicators is carried out firstly according to hierarchical levels (strategic or operational) and then according to a functional breakdown or by business process activities. Each section consists of identifying the management concerns, the measurable objectives if possible and finally the progress or result indicators to verify the achievement of these objectives.
Step 3	design of the indicators and the dashboard (the visual representation). This step consists of defining in detail the indicators chosen in step two. For each indicator, an indicator sheet is completed, specifying the unit of measurement, the calculation, the breakdowns, the comparisons, the data sources, as well as examples of graphic representation and possible interpretations. And finally, after the documentation of all the selected indicators, they will be arranged in a panorama so as to form the complete prototype of the scoreboard.
Step 4	the computerisation and realisation of the dashboard production system, this is the stage where the choice of the software package and the platform to be used for the dashboard must be determined so that it is compatible with the existing information systems.
Step 5	the implementation of the dashboard allows the prototype developed to be tested in the field as a pilot project and to make the necessary changes before moving on to full deployment.

Voyer's involvement in various dashboard projects in the health sector and in public administration in general has enabled him to build up a bank of general indicators which he suggests should be used as a source of inspiration during the indicator identification phase.

3.4.3 Boix and Feminier's approach (2003)

Boix and Feminier (2003) propose a methodology for designing a scoreboard in the form of thematic sheets that can guide the scoreboard project. This approach consists of five essential phases that can be summarised in the table below (Table 3.3).

Table 3.3 Phases of dashboard development according to Boix, D. and Feminier, B. (2003)

Phase 1	Stage 1	Identify the missions of the system
Setting control objectives	Step 2	Identify changes in the system context
	Step 3	Defining control priorities
	Step 4	Define control objectives
Phase 2	Stage 1	Identify possible indicators
Defining indicators	Step 2	Choosing indicators
	Step 3	Define precisely the indicators selected
	Step 4	Validate the coherence of the control system
Phase 3	Stage 1	Identify the control history
Defining control indices	Step 2	Defining the tolerance range
	Step 3	Defining data sources
	Step 4	Formalise the dictionary of indicators
Phase 4	Stage 1	Defining the model
Formatting of the	Step 2	Develop a prototype
on board and prototyping	Step 3	Testing the prototype
	Step 4	Validate the results
Phase 5	Stage 1	Collecting data
Operation of the dashboard and	Step 2	Interpreting the results of syntheses
	Step 3	Define corrective actions
maintenance of its evolution	Step 4	Evolving the dashboard

3.3.4 Synthesis

The proposed approaches or methodologies provide broad guidelines for designing a scorecard. They give us a global vision of the main dimensions to be measured and the types of measures that should be chosen for each of the dimensions so that they can reflect the company's strategy, but they do not offer the necessary methods to know how to go about carrying out these tasks.

3.5 Supply Chain Scorecard

The American approach to dashboard design "the balanced scorecard" is a vehicle that reflects an organisation's mission and strategy in a set of objective and quantifiable measures organised into four different perspectives: financial, customer, internal processes and learning and growth. According to several authors the four perspectives of TBP are appropriate to overcome the problems related to performance assessment in the supply chain.

3.5.1 Use of the Balanced Scorecard to assess supply chain performance

According to Zimmermann and Seuring (2009), TBP has gained increasing acceptance as an instrument for implementing business strategies, and transforms them into linked performance measures, which can be extended to the evaluation of supply chain performance. Bhattacharya et al (2014) cite the following reasons for using TBP in this evaluation:

J The objectives of supply chain management (reduction of service time, response flexibility, cost reduction, new product introduction) can be measured by the internal process perspective;

J The results of supply chain management can be measured through the financial and customer axis;

J The rate of improvement in supply chain management can be measured by learning and growth perspective;

J It can be used as an information system;

J It allows you to visualise the cause and effect relationships between the different measures.

Reefke and Trocchi (2013) state that the formulation of a TBP for SCM is divided into six steps:

> Defining the MSC strategy ;

> defining the scope of application ;

> identification of environmental and social exposure ;

> determining the strategic relevance of sustainability aspects ;

> defining cause and effect relationships;

> definition of measures and indicators.

For Bhagwat and Sharma (2007a), this process involves :

> creating awareness for the concept of TBP in supply chain management;

> collect and analyse information on companies, companies and supply chain management strategy and potential measures related to the four perspectives;

> clear definition of objectives and specific strategic goals ;

> development of a preliminary performance measurement system ;

> receiving management comments and feedback ;

> consensus on the system that will be used by the organisation, and presentation of the system to all stakeholders.

According to Rajesh et al (2012), TBP is still out of reach for most small and medium-sized enterprises, as its development requires a lot of skill and management expertise, time and money expenditure. For Barber (2008), criticisms of TBP and its many applications and developments indicate that people and suppliers are excluded, regulations and competitive environments are ignored as well as environmental and social aspects of the industry.

According to Reefke and Trocchi (2013), environmental and social aspects can be integrated into the four perspectives by setting strategic priorities that influence the formulation of the respective targets, measures and indicators, representing important strategic factors that cannot otherwise be sufficiently represented by integration into the four standard TBP perspectives.

3.5.2. Limitations identified in the literature

Naini, Aliahmadi and Jafari-Eskandari (2011) argue that there are some limitations to TBP for supply chain management such as does not take into account the relationship of cause and effect over time, does not provide mechanisms for selecting the best performance measures, does not define value chains in strategic operations, and is not dynamic enough for online monitoring.

The literature review by Agami, Saleh and Rasmy (2012) points out that most existing supply chain performance measurement systems are rigid and lack continuous improvement. In an attempt to fill this gap, the authors propose a dynamic, continuous and hybrid system framework that integrates the system under study, strategic planning, TBP, the SCOR model, theory of constraints, process thinking, optimization and structure analysis into a cohesive approach to improve supply chain performance.

According to Xian, Qiu and Zhang (2013), although the measurement of supply chain management performance can be studied as a TBP, such an approach is not effective for company-level assessment in that many measures can also be influenced by other business activities.

3.6 Bibliographic synthesis on the design of a scoreboard and positioning

In the recent literature, we find several contributions in the field of dashboarding. We present below the recent works in this field, the tools used, the field of study and the industrial context.

Table 3.4 Summary of work on the implementation of a TDB

Studies	Objectives	Proposed tools or approaches	Field of study	Context
Fana Rasolofo-Distler, 2009	Design of the CSR dashboard system	Combination of TBP and OVAR	Social enterprise for housing	French
Mabrouk AIB et al, 2010	Designing a strategic dashboard	The Balanced Scorecard	Application to the upstream activity of an oil company	Algerian
Frederic Bonvoisin, 2011	Development of a system of performance indicators (SIP)	Development of a four-step methodology for the design of the operating theatre scorecard	Hospital	French
Azzouz Elhamma, 2011	Impact of the size on the content of dashboards in Moroccan companies	Empirical study on 62 Moroccan companies (SMEs and large companies) on the use of a balanced TDB (TBP)	Management control	**Moroccan**
Frederic Juglaret, 2012	Indicators and dashboards for risk prevention	a balanced scorecard	Health and Safety at Work	French
Joelle Morana, Gilles Pinardi, 2012	Designing a sustainable scorecard	TDB which includes the three categories of development	Urban logistics	French
	a delivery pooling system	It includes seven main indicators (4 economic, 1 environmental and 2 social/social).		
Fatima Ouzayd et al ,2012	the development of a logistics TDB	the ASCIS modelling and simulation approach	Hospital	**Moroccan**
Houda zian, 2013	To propose an explanatory model of TDB practices in Moroccan SMEs.	Analysis of TDB practices in SMEs in the literature using a field survey	Management control	**Moroccan**
Imane Ibn farrouk, 2013	Design of a TDB of the drug supply chain	Proposal for an "OPRI" approach Objectives, parameters, risk, indicators. Based on the SCOR model	Hospital sector	**Moroccan**
Mahrat Fahd et al, 2014	Design of a system of industrial safety performance indicators	Proposal of a combinatorial method (TBP, ECOGRAI...)	Occupational Safety and Health (OSH)	**Moroccan**

The review of the literature for the design of a dashboard has enabled us to draw some remarks :

■ The literature targeting the management of a company's supply chain by dashboard is almost absent. Recent works analyse the practices of TDBs and their content in terms of the type of indicators used in the framework of management control. Indeed, according to the survey conducted by (Azzouz, 2011) on the indicators included in the dashboards of 62 companies established in Morocco and their impact on overall performance, the author stresses that the performance measured by the steering systems is mainly financial. The majority of companies make little use of indicators relating to customers and internal processes. As for the "organisational learning and innovation" axis, it is hardly represented in the dashboards. The author confirms that balancing the scorecards has a positive and statistically significant impact on performance. The results of this survey show that the more balanced the dashboard, the better the performance of the company.

■ The literature review shows the importance of TBP as a supply chain performance measurement system. And several authors have proposed steps for successful implementation of such a system.

In this context, our contribution satisfies this need by proposing a hybrid methodology to design a supply chain dashboard adapted to the industrial context that integrates the TBP and the SCOR model. In this context, we have proposed a hybrid methodology to design a supply chain dashboard adapted to the industrial context that integrates the TBP and the SCOR model. Thus, the proposal of an approach for a relevant choice of indicators responding to the objectives of the supply chain in order to eliminate the possible deficiencies altering the good functioning of the chain.

Indeed, the above-mentioned approaches do not address the company's supply chain. And the other proposals for developing a TBP for a supply chain only give guidelines and do not offer tools on how to implement them. Furthermore, they do not integrate risk as a performance measurement criterion and they do not provide a clear formalism for the selection of appropriate indicators.

3.7 Conclusion

The efficient management of the supply chain is inextricably linked to a constant monitoring of performance. Indeed, the supply chain is an essential component of the company's strategic approach in the quest for competitive advantage. A well-designed management dashboard composed of judiciously chosen indicators is essential.

However, its design often poses difficulties in terms of choosing indicators and meeting the expectations and needs of its different users.

In this chapter, we have discussed:

1. The concept of the dashboard and its different approaches: The French OVAR approach (the French dashboard) and the American approach (the prospective dashboard);

2. Methodologies and practical guides to dashboard design ;

3. Dashboard management in the industrial context ;

4. Review of the literature on the design of a balanced scorecard for the supply chain ;

5. A literature review of recent work on dashboard design and our position in relation to other work.

The next chapter will be devoted to the presentation of our methodology for designing a supply chain dashboard.

Chapter 4: Methodology for developing a global supply chain scorecard

4.1 Introduction

In order to develop a global supply chain dashboard, it is imperative to have a robust methodology. In this chapter, we propose an original methodology which is articulated around four phases:

- ♦♦ **Phase 1:** Description of the study area ;
- ♦♦ **Phase 2:** Global supply chain analysis and modelling;
- ♦♦ **Phase 3:** Designing a process dashboard ;
- ♦♦ **Phase 4:** Design of a global supply chain dashboard.

For each of these phases, we propose the steps leading to the achievement of their objectives and the tools used.

After presenting our methodology, we will present our empirical study on the main indicators considered important for industries in Morocco.

4.2 Methodology for the elaboration of a global dashboard of a supply chain

4.2.1 Reflection:

Supply chain performance measurement has become an activity of recognised importance, particularly due to the complex nature of the processes, usually involving several decision criteria.

Supply chain management impacts not only the overall performance of the company, but also the competitive advantage of organisations. In supply chain management, performance evaluation aims to obtain information on activities that are not appropriate to the established objectives in order to redirect its course and also to identify opportunities for improvement.

Performance measurement should be addressed and managed systematically in an organisation; when performance is below target and requires immediate action to avoid impacting the bottom line, and when performance is above target and sets a new target.

The main benefit of a scorecard-type performance measurement system is to provide a comprehensive and timely framework of information on the performance of an organisation. Another contribution is to allow a diagnosis of the company's weaknesses and to decide when and where corrective actions are needed in order to assess the impact of these actions on performance.

While the primary purpose of a company is to make a profit and achieve good financial results, Norton and Kaplan (1996) have shown that company performance can be measured along other dimensions. Thus, the financial, customer, internal process and organisational learning dimensions combine performance drivers that have the ultimate and overall objective of improving the financial performance of the company.

In order to measure the company's performance within its supply chain, we rely on a set of performance indicators derived from the following dimensions

- Financial axis which groups together all the indicators relating to the economic performance of the supply chain;
- Customer axis, which groups together all the indicators that measure the customer's vision;

- Internal process axis which corresponds to measures monitoring the **key points of the internal supply chain** where the company has to be particularly efficient;

- Organisational learning axis which reflects the ability of the supply chain to progress and implement conditions to increase value creation.

Starting from these performance dimensions, our reflection consists of piloting the internal processes that make up the supply chain to achieve a global piloting of the chain.

For the management of processes, a global modelling by adopting the process approach seems to us indispensable. It allows for a global approach to the supply chain within the company and considers a sequence of activities from the supplier to the customer. For this purpose, we adopt the modelling approach recommended by the SCOR model.

Other than modelling, we use this model to :

- identification of process malfunctions;

- identifying risks within processes to develop new operational objectives;

- validation of indicators with our own bank of indicators.

These determinants are grouped in a global and general approach to build a dashboard by process of the chain (Naciri et al, 2015b). To present our indicators that provide information on the achievement of operational objectives, we draw on the TBP approach to classify them according to performance axes.

For the overall management of the supply chain by means of a dashboard, we follow the logic of the TBP, but the strategic objectives selected are those set out in the strategy and reviewed by the operational objectives developed at the stage of management of the chain processes.

Finally, we can say that our supply chain management thinking is a kind of integration between the SCOR model and the Balanced Scorecard in order to produce an efficient performance measurement system by coupling the top-down and the buttom-up approach.

This reflection is translated into a methodology that covers all the elements necessary for the establishment of an effective SIP and TDB. It is structured around four phases of implementation, which we will detail in the next section.

4.2.2 Methodology

4.2.2.1 Phase 1: Description of the study area

Our final objective being to elaborate a global dashboard for a logistic chain, it was necessary first of all, to obtain a maximum of information on the functioning of the company. To do this, we started by:

- *Identification of the integrated supply chain of the company*

The objective of the study of this logistics chain is to situate the company within a chain or network of actors extending from the first producer to the final customer. This study allows us to identify the physical and information flows that link these actors and to master the functioning of the company.

- *Characterisation of the company :*

- General **knowledge of** the company

- **The mission**

The mission identifies the purpose of the company, it qualitatively answers the question: "What does the organisation want to mean to its stakeholders?

This mission must serve as the basis for defining long-term objectives and is the guiding principle for the actions of managers and employees

- **The vision**

The vision offers an inspiring, dynamic image, implying a challenge for the future: "What do we want to achieve in five years? What do we want to represent then for our environment and for all stakeholders?

The vision specifies how the organisation should behave in achieving its mission.

- **The strategy :**

The strategy, on the other hand, is the set of manoeuvres put in place to achieve the vision. The strategy is made up of small action plans to which dates and responsible persons are attached for the development of the necessary objectives.

Understanding the goals and objectives of the business is a step that is present in all steps of implementing a performance measurement system. Performance indicators should be developed in relation to the strategy.

- **Company processes**

Three types of processes are included in the company's mapping: management processes, support processes and operational processes (Valla, 2008).

Determining these processes will help us to identify those that are included in the company's supply chain. The figure (Figure 4.1) shows a schematic of this first phase.

Phase 1: Description of the study area

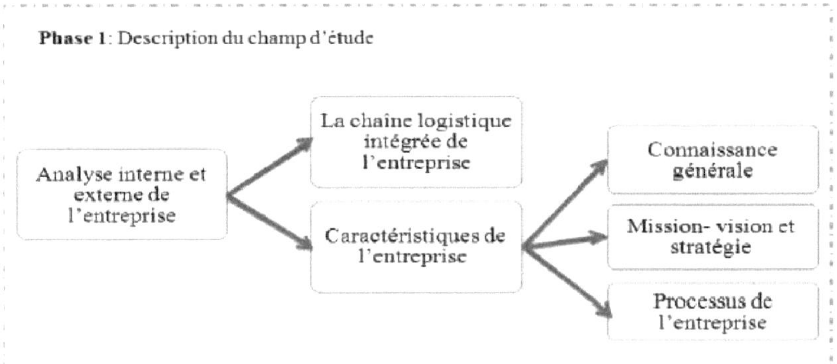

Figure 4.1 Phase 1 modelling

4.2.2.2 Phase 2: Global supply chain modelling and strategic diagnosis

The analysis and mastery of the supply chain operation is a primordial step, which allows both to carry out a relevant modelling of the supply chain, and to carry out a strategic diagnosis in order to establish the strategic objectives and the key success factors of the company.

The objective of this modelling is the description of the supply chain via processes. The main challenge is to obtain an objective, precise and clear representation of the functioning of the chain. To do this, we opt for the SCOR model.

For strategic diagnosis we use the **SWOT** (Strengths - Weaknesses - Opportunities - Threats) or MOFF (Threats - Opportunities - Strengths - Weaknesses) method, which is a very useful tool in the strategic diagnosis phase. It has the advantage of summarising the strengths and weaknesses of a company in relation to the opportunities and threats generated by its environment (Figure 4.2).

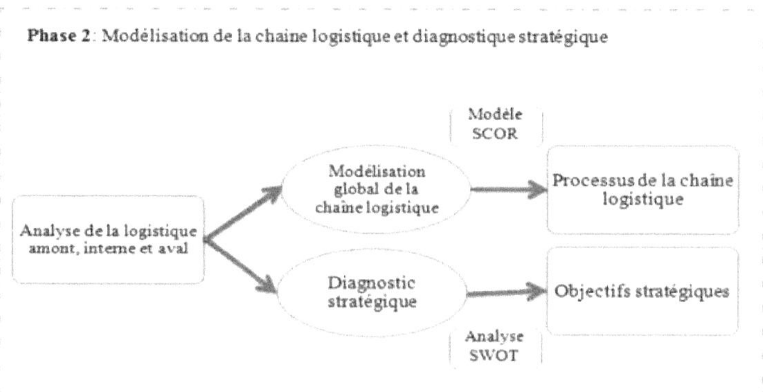

Figure 4.2 Phase 2 moderation

4.2.2.3 Phase 3: Design of a process dashboard

The performance of internal logistics flows can represent significant financial stakes. However, the organisational principles and rules for managing these flows are often far from optimal. The lack of visibility on the performance of these flows leads to a lack of management and a loss of efficiency.

In this sense, we propose an approach to measure the performance of operational processes in order to pilot the internal logistics flows. This approach is based on the analysis of dysfunctions and risks affecting their proper functioning in order to build an efficient system of indicators.

The different steps of our approach are detailed in the next sub-sections.

4.2.2.3.1 Functional analysis :

The aim is to carry out a complete analysis of each of the processes in order to identify the dysfunctions that penalise the performance of the supply chain. The objective is to obtain an objective and exhaustive identification of the problems and their validation. To analyse the process, we propose :

■ Process identification and description: To facilitate the work in the subsequent steps, the analysis of a logistics system is necessary for a detailed characterisation.

■ Process categorisation: In accordance with the second level of the SCOR model the major groups are subdivided into process categories, which are: four to planning, three to procurement, four to distribution, six to return (three to procurement and three to distribution), and five to support. The three categories are subdivided into: stock, demand, and design to order, but distribution has a fourth category, which is the sale of the retail product. Return in turn has three categories: Defective product,

product for general maintenance and repair, and surplus product.

- Establishing the level of process detail: At this stage, it should represent processes in more detail. This is achieved by breaking down the categories outlined in the previous step into process elements.

4.2.2.3.2 Risk identification and target setting

After the analysis, we carry out a process failure study. We perform this study by comparing the current processes of the company with the (ideal) processes of SCOR. This study allows us to detect both the dysfunctions and the risks involved in the process.

As for the setting of operational objectives, they are determined by the collective work of a working team composed of the company director and the process manager. The required objectives include those that are stated in the company's strategy and those that address the risks affecting the proper functioning of the process.

Figure 4.3 Improved operational objectives

4.2.2.3.3 Choice of performance indicators

The main purpose of this research work is the construction of a methodology for the elaboration of a supply chain dashboard adapted to the industrial context. This objective requires a relevant choice of indicators which guarantees the effectiveness of a good performance measurement system.

In this framework, we have carried out, first, a synthesis of indicators of the main processes of the supply chain that we have classified according to the four classic dimensions of the Balanced Scorecard. Then, we subjected these indicators to a selection through a questionnaire survey among 266 industrial companies (Naciri et al, 2015a). As a result, we designed a database of performance indicators deemed important to feed our dashboard.

Since the SCOR benchmark proposes performance indicators for the supply chain processes, we propose to intersect these indicators with those from our database. The "filter" indicators are the indicators that are common to both sources and that will subsequently form a reference for the selection of our own indicators.

The added value of this technique is to build a base of reliable and relevant indicators and, at the same time, approved by the global benchmark for performance measurement.

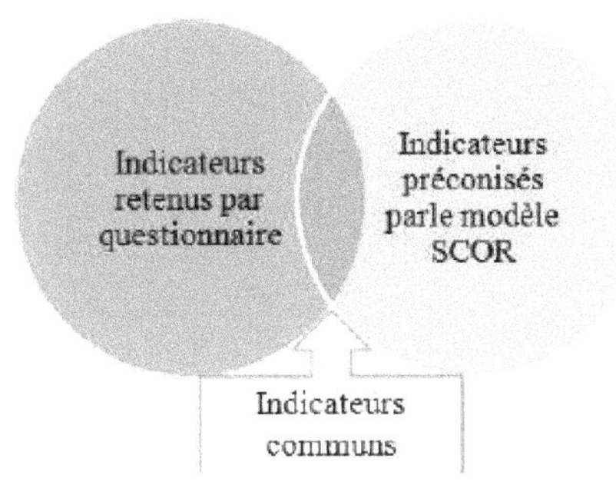

Figure 4.4 Sources for obtaining indicators

To build our system of process indicators, we start from the operational objectives per perspective of the TBP. We then break them down into key process factors (CPF). We then propose the indicator that corresponds to this KPF by referring to our "filter" indicators.

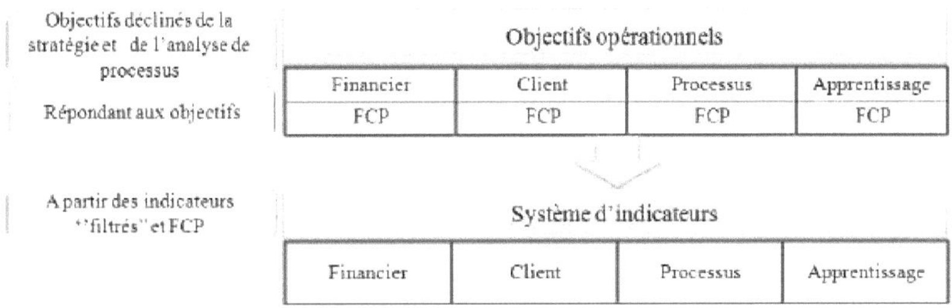

Figure 4.5 SIP identification process

4.2.2.3.4 Development of a process dashboard

Finally, we group the selected indicators into a dashboard, specifying the calculation method, the performance target, the actors involved and the action plans. Figure 4.6 models the steps in this phase.

Phase 3: Design of a process TDB

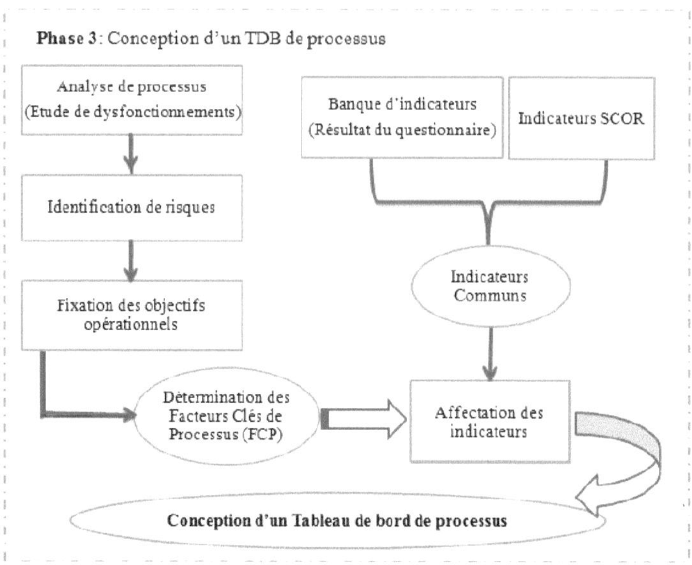

Figure 4.6 Phase 3 modelling

4.2.2.4. Phase 4: Designing a global supply chain scorecard

The evolution of company strategies requires the definition of new instruments for measuring global performance. (Dixon *et al.* 1990) propose a pyramidal representation of the organisation based on a triptych of Strategy, Actions, Measures.

The top of the pyramid reflects the strategic vision of the company. Consequently, the *Actions* to be implemented and the Performance *Measures* must be adapted to the defined strategy. The achievement of the activities, constituting the base of the pyramid, and the intermediate objectives are the guarantee of the achievement of the strategic objectives.

Given their importance, we identify the strategic orientations of the company by using the SWOT method. These objectives are then broken down into operational objectives (OO) for each process in the company's supply chain. In the third phase, the operational objectives are reviewed through a process analysis.

Seeking to achieve a balance between the different categories of objectives, and in order to converge the expectations/needs of the leaders with those of the actors in the field, we propose to carry out a revision and validation of the strategic objectives by those of the operational ones. This update seems to us to be crucial to take into account the operational constraints in the formalisation of the strategy.

To create our global dashboard, we draw on the global, multi-criteria performance philosophy developed by Kaplan and Norton. And to feed it, we will suggest global indicators for the Balanced Scorecard axes that respond to the defined strategic objectives. These indicators are chosen on the basis of our literature review (Naciri et al, 2014b) and the level 1 indicators proposed by the SCOR model. Figure 4.7 summarises our approach.

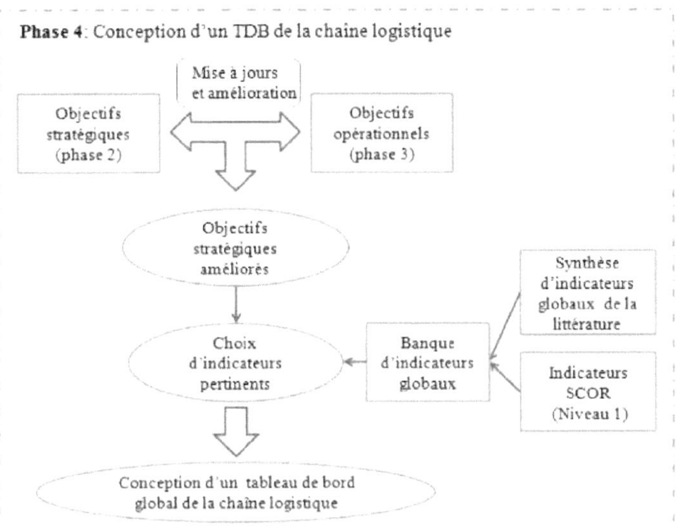

Figure 4.7 Phase 4 modelling

4.2.3 Synthesis:

Our methodology for the design of a global supply chain dashboard is a contribution that couples :

- l top-down approach which is effective in ensuring that the strategy is implemented as defined in the top management and takes little account of the expectations and ambitions of the actors on the ground,

- And the buttom- up approach involving the process actors and taking into account the operational constraints.

Knowing that the choice of truly relevant indicators is the key to the success of any steering project, our methodology proposes an effective method for choosing and constructing truly relevant indicators that contribute to real decision-making. Figure 4.8 illustrates the overall modelling of our methodology.

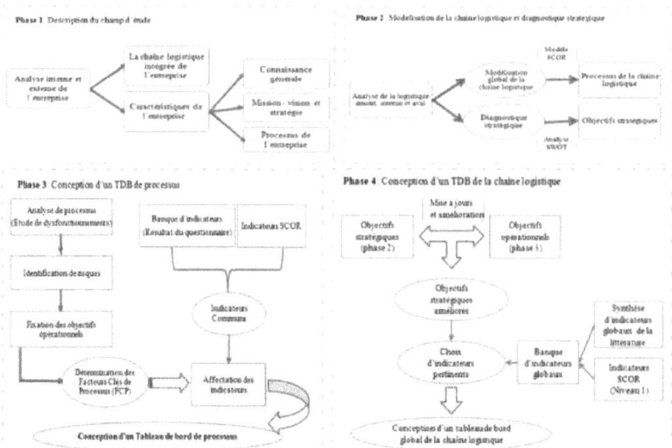

Figure 4.8 Different phases of our supply chain management methodology

4.3 The implementation of the empirical study

Now that we have detailed our methodology, we will present the empirical study to select the performance indicators that are adapted to the industrial context and that are derived from the dimensions of the Balanced Scorecard.

In this work, we study the degree of importance of performance indicators of companies located in different geographical areas. In this case, the questionnaire survey is a technique that meets this objective (Igalens and Roussel, 1998; Newsted et al., 1998). A questionnaire is a tool that "makes it possible to question individuals directly by defining in advance, through a qualitative approach, the modalities of response through so-called closed questions" (Baumard et al., 2003).

This section describes in detail the sample selection phase, the design and delivery of the questionnaires, and the data collection.

4.3.1 Selection of the sample

Before starting the field research, it is necessary to define the population to which the survey is addressed. For this purpose, we contacted the Chamber of Commerce to obtain the addresses of companies located in Morocco. Two criteria were used to define the companies constituting the base population.

- Certification: companies must be certified to at least ISO 9001.
- Geographical location: the study covers 14 regions composed of 59 Moroccan cities.

After having specified the criteria defining the population of our study, we have to choose a sample among the companies present in the list in our possession. This leads us to specify the sampling method.

From a theoretical point of view, two types of methods can be distinguished:

- Probabilistic: the sample is obtained by a random drawing procedure in which each element of the population has a known, non-zero probability of being drawn;

60

- **Empirical**: in this case, the constitution of the sample is the result of a reasoned choice, the companies are selected by applying certain rules or choice criteria aimed at making the sample resemble the population from which it is drawn (Evrard and Lemaire, 1976).

Because of the criteria defined above, and also because of the objective of our study, both types of methods were necessary for the definition of our sample and the concrete identification of the companies to which our questionnaires will be sent.

The empirical method allowed us, in a first step, to extract from our list the companies certified at least ISO 9001 located in the different regions of Morocco.

Then the probabilistic method allowed us to randomly draw our base sample. Table 4.1 shows the distribution of the mailings by region.

Table 4.1 Distribution of shipments by company size and region (Naciri et al, 2015a).

Region	City	By post	By electronic means	Face to face	Total
CHAOUIA-OUARDIGHA	BEN AHMED	1			1
	BEN SLIMANE	1	5		6
	BERRECHID	5			5
	KHOURIBGA	3	4		7
	SETTAT	1	2		3
DOUKALA-ABDA	AZILAL		2		2
	BENI-MELLAL	1	3		4
	EL JADIDA		3		3
	SAFI	4	2		6
	TADLA		2		2
FES-BOULMANE	BOULEMANE	1	2		3
	FES	10	4	11	25
	SEFROU		1		1
	MOULAY YACOUB				0
GHARB-CHRARDA-BENI HSSEN	KENITRA	5	4		9
	SIDI KACEM		1		1
LARGE CASABLANCA	CASABLANCA	12	20	3	35
	MOHAMMADIA	7	10		17
	MEDIOUNA	1	2		3
	NOUACEUR		1		1
GUELMIM ES-SEMARA	ES-SEMARA		1		1
	GUELMIM	2	2		4
	TAN-TAN	1	2		3
	TATA				0
LAAYOUNE-BOUJDOURSAKIA HAMRA	BOUJDOUR	1			1
	LAAYOUNE		1		1
MARRAKECH-TENSIFT-AL HAOUZ	AL HAOUZ	1	2		3
	CHICHAOUA		3		3
	EL KELAA SRAGHNA				0
	ESSAOUIRA	1			1
	MARRAKECH	4	5		9

Region	City	By post	By electronic means	Face to face	N Total
MEKNES-TAFILALET	MEKNES	4	5		9
	EL HAJEB				0
	ERRACHIDIA	1	1		2
	IFRANE				0
	KHENIFRA		2		2
RABAT-SALE-ZEMMOUR	KHEMISSET		1		1
	RABAT	6	5	9	20
	SALE	2	4		6
	SKHIRATE- TEMARA	11			11
EASTERN REGION	BERKANE	1	1		2
	FIGUIG		1		1
	JRADA		1		1
	NADOR	2			2
	OUJDA	3	1		4
	TAOURIRT		1		1

SOUSS MASSA DRAA	AGADIR	2	4		6
	AIT MELLOUL	5	3	5	13
	OUARZAZATE		1		1
	TAROUDANNT	1			1
	TIZNIT		1		1
	ZAGORA		1		1
TANGER-TETOUAN	CHEFCHAOUEN	1			1
	LARACHE		2		2
	TANGER	4	3	3	10
	TETOUAN	1	2		3
TAZA-AL HOCEIMATAOUNATE	AL HOCEIMA	2	1		3
	TAOUNATE	1			1
	TAZA	1			1
		110	125	31	266

After presenting our study sample, the next section presents the method of designing the questionnaires.

4.3.2 Designing the questionnaires

As we have specified before, we seek in this work to study the importance of indicators in the management of the performance of companies and to try to propose a typology (Naciri et al, 2015a).

In order to meet these two objectives, the development of the questionnaires was carried out in three stages: a review of the literature, the testing of the questionnaire with a group of four people (a university professor in research methodology, a quality manager and a logistics manager) and the testing of the questionnaire with three company managers.

-   The literature review: to gather the performance indicators existing in the literature for the main physical processes of a supply chain (procurement, production, delivery), we have collected them based on the works of (Colin and Pache, 1988; Michel et al. 1989; Florence and Laurent, 2007; Demarcheiso, 2010; Morana and Pinardi 2003; Valentine et al. 2007; Gunasekaran and Kobu 2007; Griffis et alii 2007; KPI's 2015)

-   Questionnaire testing: Questionnaire testing is an important aspect of survey research quality (Pinsonneault and Kraemer, 1993). It has several objectives: it allows us to test the form of the questions and their order, to check the comprehension of the respondents, to examine the relevance of the proposed response methods (Baumard et al, 2003) and finally to check the required response time. In our work, we tested the questionnaire with a procurement manager, a production manager and a logistics manager who are prospective respondents to the survey (Van der Stede et al, 2005).

This group checked that the questions were clear and well understood. It also assessed the length of the questionnaire: a response time of between 10 and 15 minutes was set.

Two important distinctions can be made in the development of questions: open and closed questions.

■   **Open questions :**

These are questions that give the interviewee the opportunity to express themselves in several sentences. They allow for deeper questioning and bring out unexpected insights into what is being sought. However, their disadvantage is that they are time consuming and difficult to code.

■   **Closed questions :**

These are questions that offer specific answers proposed by the researcher. They have the advantage of facilitating responses, their coding and analysis.

Given the ease with which they can be used in the conduct of the survey, we have opted for closed questions in our questionnaire. On the other hand, the use of open-ended questions in this type of research seems inappropriate.

The first part includes questions on the characteristics of the company and the interviewee. The second part concerns the degree of importance of the selected performance indicators.

-    Part 1: The first part of the questionnaire consists of the closed questions, which are general questions about the companies and the respondent; they collect data about the respondents, the type, size and sector of activity of the companies in our sample.

-    Part Two: The second part concerns the importance of the performance indicators and is composed of three-choice questions. We have associated with each indicator a grid of answers which contains three columns corresponding to the following evaluations: "important", "rather important" and "not important". And we have proposed a classification of the indicators, according to the principles of Norton and Kaplan's (1996) Balanced Scorecard. (Appendix)

Once the questionnaire was completed, we wrote a covering letter that specified the purpose, objective, interest of the survey and the return date. In addition, to reassure respondents, we insisted on guaranteeing the anonymity and confidentiality of responses.

After presenting our method of designing the questionnaires, we present below their method of delivery.

4.3.3 Sending out the questionnaires

Because the objective of the research is to measure the degree of importance of logistics performance indicators, the need to collect a sufficient amount of data appeared important. For this purpose, the questionnaire survey was sent out by post, electronically and face-to-face.

4.3.3.1 Questionnaire sent by post and electronically

This is the first stage of sending out our questionnaire. We sent out 110 questionnaires by post.

In turn, the mail survey offers the following advantages: -It minimises the impact of the researcher and reduces response bias;

-    This gives the survey more credibility.

A major drawback of the postal survey is the low response rate. To overcome this shortcoming, we phoned the target companies once or several times to improve our response rate.

In this first stage, we sent 126 questionnaires electronically, which has several advantages, including

-    Email is a quick way to send questionnaires;

-    Hundreds can be sent in a matter of seconds;

-    The shipping cost is almost zero.

In this first stage, we collected 16 usable questionnaires by post and 13 by e-mail, i.e. an actual response rate of 14.54% for the post and 10.3% for the e-mail (Naciri et al, 2015a). As these rates were low, we had to go through a second stage: sending out face-to-face questionnaires.

4.3.3.2 Face-to-face questionnaire

One of the main advantages of this type of interview is that it offers more opportunities to assess the interviewee's understanding and interpretation of questions, as well as to clarify any ambiguity about the meaning of a question or answer. During an interview, it is also possible to show interviewees documents or objects and ask for their reaction.

The disadvantages of this type include

- The presence of the interviewer, which can influence the answers given by the survey;
- The cost of travel.

In this second stage, we consulted 31 companies on the spot, which enabled us to collect data from 11 companies that agreed to receive us, i.e. an actual response rate of 35.48%.

In total, of the 266 questionnaires sent out, 42 were returned complete, giving an initial response rate of 15.78%. Of these 42 responses, 12 questionnaires were not usable because of missing data, and the actual response rate was therefore 11.2%. In the end, data from 30 companies could be processed (Naciri et al, 2015a).

4.3.4 Analysis and evaluation of results

Based on the LAVINA questionnaire (Lavina, 1993), each response is assigned a weighting coefficient: 1 - 0.5 - 0. The choice of indicators according to each axis consists in calculating the sum of the points obtained according to the three columns (Naciri et al, 2014a)

We calculated the sum of the points obtained for each indicator according to the weighting coefficients, as well as their percentage of importance, and we chose the indicators that reached or exceeded 50% of importance for each process.

Tables (4.2, 4.3, 4.4) represent the results of our analysis method for the three main supply chain processes.

Table 4.2 Results and scores for each supply process axis (Naciri et al, 2015a).

Procurement process		Important	Rather important	Not important	Amounts	of importance
FINANCIAL ASPECTS						
1	-Price reduction purchase price compared to historical price	22	5	3	24,5	82%
2	- Service cost / purchase turnover managed by the service	20	2	8	21	70%
3	- Service cost/savings generated by the service.	26	4	0	28	93%
4	- Average cost of procurement of a order	29	1	0	29,5	98%
5	-Increase in Deadlines payment supplier	18	2	10	19	63%
6	- Average value of an order	13	5	12	15,5	52%
7	-Annual purchase value per supplier	27	3	0	28,5	95%
CUSTOMER FOCUS						
8	- Satisfaction rate	27	3	0	28,5	95%
9	- Cumulative number of days	28	2	0	29	97%

		Important	Rather important	Not important	Amounts	of importance
	late / number of late deliveries					
10	- Actions affecting customer loyalty	23	4	3	25	83%
INTERNAL PROCESSES						
11	- Average processing time of a Purchase request	28	2	0	29	97%
12	Stock coverage rate	28	1	1	28.5	95%
13	- No. of non-compliant batches / No. of refused batches	22	5	3	24,5	82%
14	- Number of orders refused and in progress	19	8	3	23	77%
15	- No. of active suppliers monitored	28	1	1	28,5	95%
16	- Stock rotation by product type	28	2	0	29	97%
17	- Rejection rate due to quality defects	25	3	2	26,5	88%
18	- Average processing time of a Purchase request	15	8	7	19	63%
19	- Differences between quantities rejected and quantities ordered	14	4	12	16	53%
ORGANISATIONAL LEARNING FOCUS						
20	- Purchase turnover covered by the service / total purchase turnover	19	9	2	23,5	78%
21	- Absenteeism rate	14	5	11	16,5	55%
22	- Number of hours of training	28	1	1	28,5	95%
23	- Purchase turnover / headcount	17	5	8	19,5	65%
24	- Progress bonus	22	5	3	24,5	82%

Interpretation :

From the 70 proposed indicators, we have selected 24 indicators that are considered important in the

procurement process.

Table 4.3 Results and scores for each production process axis (Naciri et al, 2015a).

Production process		Important	Rather important	Not important	Amounts	of importance
FINANCIAL ASPECTS						
1	Production cost vs last year vs budget	19	9	2	23,5	78%
2	Production cost ^ Cost of sales	20	2	8	21	70%
3	Costs associated with machine shutdown	25	3	2	26,5	88%
4	Total production cost ^ total number of units produced	22	5	3	24,5	82%
5	Cost of product defects due to raw material quality ^ Total cost of defects	18	2	10	19	63%
6	Cost of damaged products due to staff errors ^ total cost of damaged products	13	5	12	15,5	52%
CUSTOMER FOCUS						
7	- Satisfaction rate	27	3	0	28,5	95%
8	Number of customer orders per day (units / day) ^ Number of minutes worked per day (minutes / day)	23	4	3	25	83%
INTERNAL PROCESSES						

9	Number of defects produced due to raw material quality - total number of defects	28	2	0	29	97%
10	Actual production rate $^\wedge$ target production rate	22	5	3	24,5	82%
11	Standard cycle time $^\wedge$ actual cycle time	18	2	10	19	63%
12	Actual cycle time $^\wedge$ ideal cycle time (minimum cycle time)	28	1	1	28,5	95%
13	(Actual output - Rejected output)/ Actual output	28	2	0	29	97%
14	Downtime for corrective maintenance	25	3	2	26,5	88%
15	Loss of time due to return to production x productivity x price	28	2	0	29	97%
16	Time to return to production $^\wedge$ number of pieces returned to production	14	4	12	16	53%
17	Number of defects $^\wedge$ Product size	22	5	3	24,5	82%
18	Number of defects $^\wedge$ number of units produced	19	8	3	23	77%
19	Number of late finished Production Orders $^\wedge$ total number of Production Orders	15	8	7	19	63%
20	Availability rate x Performance rate x Quality rate	14	5	11	16,5	55%
ORGANISATIONAL LEARNING FOCUS						
21	Production stoppage due to lack of staff training / Total production stoppages	26	4	0	28	93%
22	No. of subscriptions to technical journals / databases	29	1	0	29,5	98%
23	Total value produced $^\wedge$ Number of employees	28	1	1	28,5	95%

Interpretation :

From the 78 proposed indicators, we have selected 25 indicators that are considered important in the

production process.

Table 4.4 Results and scores for each Delivery Process axis (Naciri et al, 2015a).

Delivery process		Important	Rather important	Not important	Amounts	of importance
FINANCIAL ASPECTS						
1	Transport cost $^\wedge$ Cost of sales	25	5	0	27,5	91%
2	Cost of sales = Opening stock + Purchases of goods - Closing stock	18	2	10	19	63%
3	Transport cost under drafts $^\wedge$ Total transport cost	19	1	10	19,5	65%
4	Cost of rental or depreciation of trucks	16	8	6	20	66%
5	Shipping cost Product	27	2	1	28	93%
6	Total transport cost	26	4	0	28	93%
CUSTOMER FOCUS						
7	Number of customer orders delivered per day per FTE	24	3	3	25,5	85%
INTERNAL PROCESSES						
8	Annual number of deliveries (or tons, volumes of books...)	22	5	3	24,5	82%
9	The time associated with receiving, entering and	20	10	0	25	83%

	validating a customer order.					
10	Capacity used (m3)^ capacity available (m3) during the same period	21	7	2	24,5	82%
11	Delivery variances	28	1	1	28,5	95%
12	Time of execution of orders,	22	6	2	25	83%
13	Number of deliveries per hour ^ Total number of deliveries during the same period	17	10	3	22	73%
14	% Orders delivered in full,	29	0	1	29	97%
15	The average time associated with the shipment of products	15	3	12	16,5	55%
ORGANISATIONAL LEARNING FOCUS						
16	Number of hours of training	16	8	6	20	66%
17	Number of drivers	18	2	10	19	63%
18	Absenteeism rate	20	3	7	21,5	72%

Interpretation :

From the 33 proposed indicators, we have selected 18 indicators that are considered important for the delivery process.

4.3.5 Summary

The empirical study carried out in the framework of this work allowed us to select the most important indicators for the physical processes of a supply chain. The bank of indicators resulting from this work will serve as a source, facilitating the choice of indicators for companies aiming to measure the performance of their supply chains.

4.4 Conclusion

In this chapter, we have presented our method of steering a supply chain using a dashboard, which we have divided into four main phases:

- ◆◆ Phase 1: Description of the field of study

This phase consists of presenting :

- The integrated supply chain of the company ;

- Characteristics of the company :

■ general knowledge;

■ mission - vision - strategy ;

■ business processes.

◆◆◆ Phase 2: Global analysis and modelling

This phase consists of analysing the inbound - internal and outbound logistics, carrying out a global modelling of the studied supply chain and carrying out a SWOT analysis to highlight the strategic orientations of the company.

◆◆◆ Phase 3: Design of a process dashboard

We have proposed a general and rigorous approach for the development of a dashboard for the main processes based on the following steps:

> Step 1: Analysis of the functioning of the process ;

> Step 2: Risk identification and target setting;

> Step 3: Choice of indicators ;

> Step 4: Development of a process dashboard ;

♦♦♦ Phase 4: Design of a global dashboard

To achieve this objective, we draw on the Balanced Scorecard approach to global performance measurement. Being "top-down", this approach is based on the assignment of performance indicators to key success factors derived from strategic objectives.

In order to converge the expectations/needs of the managers with those of the actors on the ground, we propose to carry out a revision and validation of the strategic objectives by those of the operational ones. This update seemed to us crucial to take into account the operational constraints in the formalisation of the strategy.

We presented the result of our empirical study, which is a questionnaire survey conducted among 266 Moroccan companies, with the aim of grouping the performance indicators deemed useful for the main processes of the supply chain. This grouping is a reference for the choice of relevant indicators in the phase of realization of a process dashboard.

The following chapter will be reserved for the application of the first two phases of our approach to a small and medium-sized enterprise (SME) specialising in the packaging and marketing of fruit and vegetables for export.

Part 3

Industrial application of the methodology for the elaboration of a global dashboard of a supply chain

The third part of our thesis includes three chapters. These chapters will be reserved for the application and the validation of the different phases of our methodology of elaboration of a global dashboard of a logistic chain of a medium-sized company, which belongs to a critical sector of activity in Morocco.

Chapter 5: Analysis, modelling and strategic diagnosis of the company's supply chain

5.1 . Introduction

This chapter is devoted to the application of the first two phases of our methodology to an industrial case. For this purpose, we have chosen a small and medium-sized agri-food company (SME) operating in the fruit and vegetable export sector.

In this chapter, we will discuss the interest in our industrial case and we will apply the first two phases of our methodology detailed in the previous chapter, i.e. we will :

1. Present our scope by defining the actors of the integrated supply chain of the company and the scope of the study which is the main actor of this chain. Then we identify the company by addressing its mission, vision, strategy and processes.

2. Analyse the upstream, internal and downstream logistics of the company's supply chain, and carry out a modelling of the chain. And finish this phase with a strategic diagnosis using SWOT analysis which will help us to highlight the strategic orientations.

5.2 The focus is on the industrial case:

We have chosen to apply our supply chain management methodology to a small and medium-sized company for the following reasons:

> Several researchers have been concerned with performance measurement in large companies, which generally have a lot of resources to develop and apply tools to improve management and performance. For SMEs, however, such research is still rare.

> The SME has several advantages that allow the success of a project to manage and measure performance:

J It generally has a simple and flexible structure that allows it to be reactive to any change in the environment;

J This organisation has low structural costs which can give it a competitive advantage over the large company;

J Because hierarchical levels are often very low, decision-making processes are faster to deal with business-related problems.

Information is also circulated more efficiently, even if it is informal;

J In SMEs, employees can be more motivated. Indeed, they can feel more involved in the sustainability of the SME.

Thus, the choice of the industrial sector is justified by the following reasons:

> The fruit and vegetable sector has long been considered as a lever of the economy and a priority axis of development for which we have been able to exploit its comparative advantages which allow us to position ourselves on new markets due to the progress made in production, packaging, marketing and export.

> In a context of globalisation and trade, the agri-food business has had its share of influence and

market on a world scale thanks to the exports of agricultural products present in various foreign markets which are characterised by the requirement of the quality of imported products, This is a real constraint for the actors of the sector, and reflects their evolving contribution towards performance, which constitutes a major stake for the survival of the enterprises and the maintenance of the good development of agriculture on a regional and national scale.

5.3 Application of the first two phases of the methodology

5.3.1 Phase 1: Description of the study area

In this phase we will present :

- The integrated supply chain of the company ;
- Scope of the study ;
- Company identification :
- general knowledge;
- mission - vision - strategy ;
- business processes.

5.3.1.1 The integrated supply chain: The early fruit export supply chain (ELC)

The presentation of the export supply chain (ELC) will be understood by defining the actors of the ELC. The main actors in the export chain of early fruit are : Producers, packing stations, exporting groups and distributors.

> *The producers*

The number of farmers in the region is estimated at 80,000, or 7% of the total population of the region. The vegetable farms are small, on average less than 5 ha. The farms adopt modern production systems. They use state-of-the-art technology to make the most of their profits.

> *Packaging stations*

The second step after production is the packaging of the produce. Morocco has more than 325 packaging stations for vegetable products. These stations have different production capacities. Tomato and potato are the main packaged products.

There are three types of conditioning stations:

- cooperative or commercial stations to which the producers belong;
- privately owned stations that buy the harvest from the producers;
- stations that are owned by one or more producers. They purchase early fruit from small producers quite commonly.

■ *Exporting groups*

Export groups are entities formed around one or more packing stations. Their role is to defend the interests of the associated members, to ensure the provision of logistical and transit services and to group shipments in order to unify the commercial strategy. These private export groups are either private production and export companies or cooperatives.

- ■ *The distributors*

For the national market, the sorting residues as well as the quantities rejected at the packing station are sold to intermediaries who come to the station to buy supplies. At the level of distribution for the international market, the exporting groups work with :

S Commissioners

They are the main sales channel for tomatoes in the region. They take the goods on consignment and sell on behalf of the exporters. They are paid a commission on the sales made, which varies according to the place and method of sale.

S Central purchasing offices of large-scale food retailers

Sales to central purchasing offices and large-scale food retail chains by exporting groups account for only 20% of sales. It is mainly the large national or foreign private exporting groups that work with the central purchasing offices.

5.3.1.2 Study area (The conditioning station)

Packing stations are the most crucial aspect and actor in export-oriented agriculture. Indeed, as the experts from AMI (Agribusiness Marketing Investment) point out, "*this is where the value of the production is added. According to the operators of the sector, price differences of nearly 0.15 Euro/kg between a well-calibrated tomato, uniform in colour, in variety, bearing a known brand, are easy to obtain. The operators, having realised the profitability of a good packing station, are investing more and more in this function.*

In order to comply with market regulations and requirements on the one hand, and to face increased competition on the other, fruit and vegetable packing and exporting stations are obliged to improve their production in the best conditions, within the shortest time and at the lowest cost. This is achieved by managing the performance of their internal supply chain (Naciri et al, 2015b).

Therefore, it becomes essential to be able to establish a good management of physical and information flows, to solve the problems faced, and to overcome the challenges of competition. This has led to a strong interest in this area.

Thus, our study area is a packing and marketing station for fruit and vegetables intended for export. It is a station to which the producers belong and ensures their sales either by the exporting group or directly to the final customer in the case of a special order.

5.3.1.3 Company identification

The company presentation will be structured as follows:

- ■ General knowledge of the company;
- ■ Mission, objectives, strategies ;
- ■ Business processes.

This information is gathered through an interview guide, intended for the station manager, and internal company documents.

5.3.1.3.1 General knowledge of the company :

The company specialises in the packaging, wrapping and marketing of fruit and vegetables for export. Equipped with the latest technology and qualified personnel, the fruit and vegetable company is committed to meeting the requirements of the foreign market.

5.3.1.3.2 The company's vision, mission and strategy

■ Company mission:

The company's mission could be defined as producing, packaging and marketing early fruit and vegetables, meeting the customer's requirements in terms of product conformity, deadlines and responsiveness.

■ Vision of the company:

The managers want the company to be the reference in its field thanks to the recognised satisfaction of its customers, the motivation, dedication and friendliness of its staff, and to be a model of a successful Moroccan company.

■ Company strategy:

Our company, like other SMEs, does not adopt a long-term strategy. However, since our approach is based on strategy, it is essential to want to identify it. To do this, we proposed to replace the company's strategy with its short and medium term objective.

His objective would be to increase his turnover by 3 to 5% per campaign and keep his costs stable.

5.3.1.3.3 Company processes

As the station is ISO 9000 certified, it has a process map. There are six processes in the company:

■ the maintenance process, which aims to ensure preventive maintenance of the station's equipment and infrastructure in order to ensure the proper functioning of the quality system. To ensure an adequate working environment at the bottom of the station;

■ the training process, which aims to train and retrain staff and ensure that everyone who performs tasks within the station is competent on the basis of training;

■ the customer process aims to meet customer requirements and ensure customer satisfaction;

■ the supply-packaging process aims to ensure that the supply of raw materials from the farms to the packaging plant is under control;

■ the purchasing process aims to control the purchase of any product or service that has an impact on the quality and control of all suppliers to the station through their evaluations;

■ the improvement process aims to ensure the continuous improvement of the station's quality system.

5.3. 2 Phase 2: Supply chain modelling and strategic diagnosis

5.3.2.1 Supply chain analysis :

We have divided our analysis into three stages of the supply chain: upstream logistics, internal logistics and downstream logistics.

5.3.2.1.1 Upstream logistics

The export group ensures the export programme. According to a schedule, the station carries out the production planning as well as the packaging planning. The company has little reliable and timely information to quantify the future demands of the export markets. These demands fluctuate week by week.

It often reproduces the reality of the previous years' production, corrected by the hazards that have marked the mind of the company manager (too rapid ripening of the products according to the climate of the last few years, overproduction according to real demand). The production of early fruit is assured by farms that are members of the company.

The farms have an average daily production of 120-300 tons. Tomatoes are the main product of the farms belonging to the company. It represents 80% of the total production. The transport of the production is taken care of by the packing station. Depending on the location of the farms, the transport time varies between 1 hour and 20 hours.

The quantity of goods collected depends on greenhouse production, which in turn is directly linked to weather conditions, which is a threat given the water requirement for fruit and vegetable production.

The agricultural season is only 7 to 8 months long, starting in October and ending in May, which means that the company's export activity lasts only 8 months, the other 4 months of the year are devoted to the maintenance of the station's machinery and to making decisions about the coming year.

In case of overproduction, the station management regulates the quantities received by setting a quota, which is the maximum capacity in tonnages received, it represents the capacity of the machine in addition to the capacity of the cold rooms in stock.

The reception of fruits and vegetables at the packing station is done through batches that represent a certain quantity of product corresponding for each of them to a single variety, a single plot, and a single harvest date. Once the goods are received, the station assigns a batch number which will guarantee total traceability and help identify the origin of the products (variety, plot, harvest date) throughout the fruit and vegetable production process. In order to ensure better quality and to avoid possible problems, the station carries out controls at the entrance:

- Agreage : consists in checking that the fruit and vegetables delivered by the producer comply with the specifications;

- Tonnage : to check the weight of the goods delivered;

- Verification of the conformity of the product (treatment used, time to harvest...);

- Control of the delivery note: which is nothing else than the verification of the conformity of the delivered goods with the delivery note.

Reception is one of the crucial steps in the supply chain process, as the station is supposed to select products in good condition, respecting national and international quality standards. The reception process must therefore be in line with the company's objectives and strategy.

5.3.2.1.2 In-house logistics:

In the internal analysis of the fruit and vegetable supply chain, we have identified the most details concerning the production sequence within the packing station. The product will only be ready for marketing once each stage of production has been completed.

After reception and grading, the goods are destined for pouring, washing, drying, pre-sorting, grading and colouring (which are done with the help of computers, electronic scales and cameras), and then another sorting of the goods is carried out.

After this sorting, the goods that do not comply with export standards are divided into two categories, the first of which represents products that do not comply with phytosanitary standards, in which case they are destroyed immediately. The second category concerns slightly injured products with an advanced stage of maturity that cannot withstand storage and transport, and are therefore destined for the local market.

Then the process changes depending on the market, for export the path is as follows: the boxing, then weighing the goods, a second grading, then the palletization (packaging), to finally direct the goods to the cold room before exporting.

Internal logistics has some problems that can affect the quality of the supply chain. These problems are classified as follows:

- technical: machine breakdowns sometimes last a few days while waiting for repairs;
- social: workers' strikes ;
- logistics: transport crisis (holidays, e.g. Christmas, New Year's Day, etc.).

These problems can cause serious damage that can affect the logistic process, profitability and even the image of the company with its customers, supposing a delay in production due to a breakdown of a machine, the whole production will stop, generating an accumulation of unprocessed products in the station, which will force the managers not to receive goods, and therefore automatically the goods will not be delivered within the time limit.

For the local market, wooden crates are the most commonly used for the transport of fruit and vegetables. The crates are generally of poor quality and of variable size. In addition, stacking them in trucks or during storage is often difficult. Also, overfilling of the boxes causes injuries, deformation and consequently losses of the products.

On export, packaging is well cared for and is subject to strict controls to meet the standards required by the EACCE and importing countries. Several production plants are available in Morocco, but much of the packaging, particularly cardboard, is also imported.

As far as storage capacity is concerned, the station's management has invested in the construction of more cold rooms increasing the total number to 7 cold rooms.

The diagram (Figure 5.1) links the harvesting activities in the orchards with the sorting and of packaging of exportable products carried out by the stations :

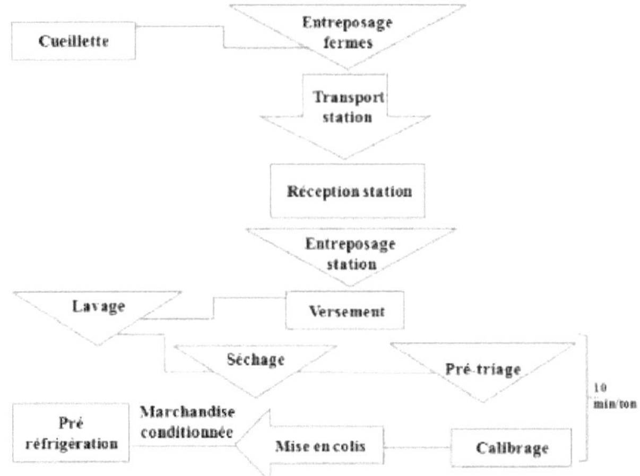

Figure 5.1 The product circuit from the orchard to cold storage

5.3.2.1.3 Downstream logistics

The goods are shipped directly after the control of the goods by the EACCE[1] , which is an organization approved by the European authorities, with as a means of delivery the road transport which is done with the help of refrigerated trailers to the exporting group, on a daily basis, and periodically as far as possible the transport by container (maritime) but this time to a destination outside the European Union, mainly to Russia.

The export group is responsible for the marketing of the primeurs on the European Union markets, mainly through its own sales offices (81% of exports).

When the company receives a special order, it opts for road transport, which meets its commitments in terms of regularity, respect for deadlines and delivery, a communication system capable of detecting and communicating with the truck where it is (fleet system), despite its advantages road transport generates significant costs for the company.

As for sea transport, it is more advantageous in terms of costs, 40% cheaper than road transport, but its use remains periodic because of its unavailability and the route is only taken once the cargo is full, which means that competitors are obliged to cooperate with the objective of gathering the most goods, in the largest number of containers, to be exported. This type of transport has many disadvantages, primarily the uncertainty of delays and bad weather at sea.

Once arrived at the destination, the trucks take care of the rest of the journey and deliver to some French supermarkets which have a contract during the whole agricultural season with the company, this type of order is very risky because the price of the products is fixed throughout the season, that is to say that insofar as there is an increase in the costs during the production, the company is obliged to sell at a loss, otherwise for the other markets, The company's salesman negotiates all the clauses of the order, i.e. the price, the deadline, the quality requested, the variables of which can be either favourable

---

[1] EACCE : the Autonomous Establishment for Export Control and Coordination

to the company provided that the quality of the product is irreproachable, the demand is higher than the supply, and the market quotas, on the other hand the negotiation turns in favour of the buyer, if the product is of average quality, the supply higher than the demand, and a bad reputation of the brand.

In case of return of merchandise, which can be frequent because the buyers when they control the merchandise, they retain the smallest detail and if they don't have 3 or 4 pallets, they can return all the cargo, in this measure the company redirects the truck to the nearest commercial platform (from the south of Perpignan or the north of Lille), in order to establish a new sorting of the merchandise in order to commercialize the less affected and to avoid heavy damages.

The supply chain analysis clearly indicates that the supply chain is one of the main concerns of packing and exporting stations. The good mastery of the supply chain will largely condition the performance of the company.

Indeed, it is the management of the supply chain that will ensure the smooth running of the station's business and guarantee the establishment of a good image among the various stakeholders.

5.3.2.2 Supply chain modelling.

The supply chain is not necessarily concurrent with the processes identified in phase 1. The first concept groups together all the processes linked to the flow of products or services from the purchase of raw materials from suppliers to the delivery of products to customers. The second concept proposes a grouping of processes that are linked by the value to the customer. The processes that integrate the supply chain processes are identified and compared to the supply chain concept in order to determine which processes are included in the supply chain in whole or in part. From this comparison, we have identified the processes that constitute the supply chain.

Next, the supply chain must be positioned in relation to these processes. For the company's management, the "supply-packaging" process is part of the supply chain and should therefore be the basis of the analysis, as should its interactions with the other processes.

After the analysis of the station's supply chain and the reflection of the working group, it was found that it contains the following five processes:

> **Planning:** this is strategic planning. This planning includes :

■ production planning by farm according to estimates prepared by the station;

■ planning the supply of fruit and vegetables according to the export group programme by variety ;

■ expedition planning ;

■ packaging planning.

> **Procurement:** this is the sourcing of raw materials and the purchase of supplies for the packaging of fruit and vegetables. This process is cascaded upstream and downstream from suppliers to customers.

> **Packaging:** The logistics chain ensures the packaging of products for export.

> **Delivery:** this is the delivery of packaged fruit and vegetables from the suppliers to the customers or the export group.

> **Returns:** returns are a very important process in the station's supply chain. It is the return of finished products that are delivered as non-conforming.

The working group validated the coverage of the entire internal supply chain of the company, from procurement from the supplier to the delivery of the customer order. In our application, the processes identified are identical to SCOR's processes. This repository has thus proved to be a very useful support tool that has brought together the different visions. Figure 5.2 gives a global modelling of our supply chain.

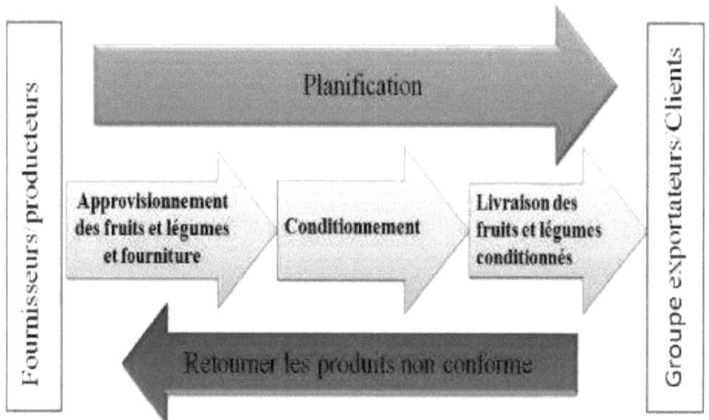

Figure 5.2 Level 1 modelling of the station supply chain (Naciri, 2015b)

These processes are somewhat different:

> The supply, packaging and delivery processes are based on the evolution of the product flow. They group together the operational tasks involved in the flow. This corresponds to more "operational" processes and to a breakdown by major functions of the physical logistics chain;

> The planning and return processes are steering processes. They deal with the flow of information and decisions for flow control.

It is believed that supply chain performance is linked to the performance of the physical processes that deliver value to the customer. These processes are :

- Procurement and the relationship with the supplier: Procurement starts with the declaration of a need and the placing of a purchase order and ends with the provision of the goods in the raw material stock;

- Packaging: the packaging process starts with the use of the goods from the machine stock until they are made available in the finished product stock;

- Delivery and customer relationship: this process starts from the finished product stock to the exporting group or directly to the final customer.

Therefore, we are interested in these three processes taking into consideration the interactions with the other processes (planning and feedback).

5.3.2.3 Strategic diagnosis

Based on the analysis of the supply chain and our research on the external environment of our company, we carried out a strategic diagnosis using the SWOT tool. The aim of this diagnosis is to take into account both internal and external factors in the strategic objectives, maximising the potential of strengths and opportunities and minimising the effects of weaknesses and threats.

5.3.2.3.1 Definition :

SWOT (Strengths - Weaknesses - Opportunities - Threats) is a very useful tool for strategic diagnosis. It has the advantage of summarising the strengths and weaknesses of a company in relation to the opportunities and threats generated by its environment.

There are two axes of analysis: the internal and the external axis:

> **Internal axis**

It identifies the current characteristics of the organisation, seen as strengths or weaknesses, depending on the activities operated.

They generally concern: human resources, production capacity, financial capacity, know-how.

Strengths: resources possessed and/or skills held that give a competitive advantage

Weaknesses: lacking in one or more key success factors or in relation to competitors.

> **External axis**

It lists elements that have a possible impact on the company.

Opportunities: The company's environment may present certain areas of potential for development. These need to be identified.

Threats: some current or future changes may have a negative impact on the company's activities.

Opportunities and threats are the external factors that create value or destroy value. They can take into account various influences in which the company cannot keep under control, (political, demographic, economic, social for example) not yet mentioned.

5.3.2.3.2 SWOT analysis

In this analysis we used the information collected in Phase 1 and documentation from the Ministry of Agriculture. The results of our analysis are summarised in a SWOT table (Table 5.1).

Table 5.1 The result of the SWOT analysis

Opportunities	Threats
■ important and diversified agroecological niches ■ favourable climate for the development of early fruit and vegetables ■ existence of a favourable transport infrastructure (air, road and sea) ■ opening of the national economy to the external market. ■ State policy to encourage agriculture: launch of the MAROC VERT project.	■ uncertain variation in the availability of fruit and vegetables, due to climatic parameters. ■ limited water resources and overexploitation of groundwater ■ risk of environmental degradation (salinity, erosion, degradation of rangelands, etc.) ■ perverse effects of the opening of the economy national external market for vulnerable farms ■ agricultural waste problems ■ new competitors ■ customs constraints including the monthly quota of exports to the European market, especially in the tomato sector ■ lack of scientific research, funding problems and quality requirements of foreign markets ■ perishability and seasonality of products

	■ the geographical distance between production and consumption areas ■ the majority of fruit and vegetable sectors are driven by the downstream sector because of the great negotiating power of large-scale distribution.
Forces	**Weaknesses**
■ quality management system ■ sophisticated and automated production system ■ very good quality product ■ commercial platforms in France that play a very important role in the marketing of products, and in reducing the risks involved in exporting, such as the return of goods ■ a multitude of markets targeted by the company Internationally (50-60%) but also a percentage (40-50%) of the production is directed to the private sector.	■ production delays caused by machine breakdowns and social problems (strikes, etc.) ■ quotas, the maximum capacity of the machine leaves managers confused during a good harvest season when production is abundant ■ a number of unused cold rooms that consume energy and additional costs ■ contracts with large retailers throughout the season at fixed prices, represents
to the national market.	a risk of selling at a loss, once production costs increase. ■ the high costs of transport ■ weak end-customer oriented commercial policy ■ difficulty in maintaining the cold chain.

5.3.2.4 Strategic and operational objectives

The performance of the supply chain can only be measured in terms of effectiveness and efficiency if the objectives are known and adapted. Based on the SWOT-analysis, the global objective of the company and the performance axes of the TBP, we define the strategic objectives of the supply chain. The strategic objectives (SO) identified are:

J Controlling costs (SO1);

J Ensure customer satisfaction (SO2):

\- To listen to the customer in order to better understand their needs today and to anticipate their needs tomorrow;

\- Meet the customer's requirements in terms of product conformity, timeliness and responsiveness.

J Maintain a high level of quality and ensure continuous improvement in order to retain an increasingly demanding clientele and to face up to very aggressive competition.

With this in mind, the company has decided to adopt a quality, safety and environmental policy (OS3) for the packaging and marketing of fruit and vegetables for all its activities and sites, based on the BRC[2] , HACCP[3] and Sustainable Development standards.

This policy is broken down into measurable objectives at the level of the company's processes and activities in order to give concrete expression to the quality guidelines on the ground, and which are reviewed regularly to ensure that they are constantly appropriate.

J Develop human resources to build a team recognised for its performance in the areas of activity (SO4).

---

[2] BRC: British organisation of food retailers
[3] HACCP: the principles of food safety (Application of FMEA analysis to the food industry)

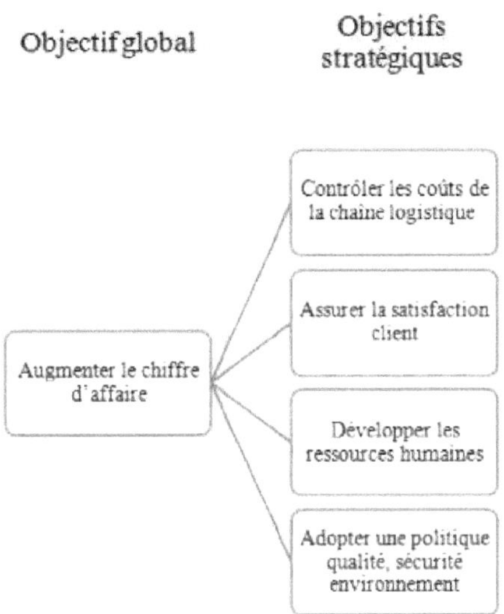

Figure 5.3 Overall objective and strategic objectives

Management by process thus implies translating the company's objectives into intermediate objectives specific to each process and then to each organisational unit that contributes to the production of added value. We thus move from strategic objectives to operational objectives. This assignment of objectives makes it possible to specify the concrete performances to be measured in an operational framework, to deduce the indicators and the values of the alert thresholds.

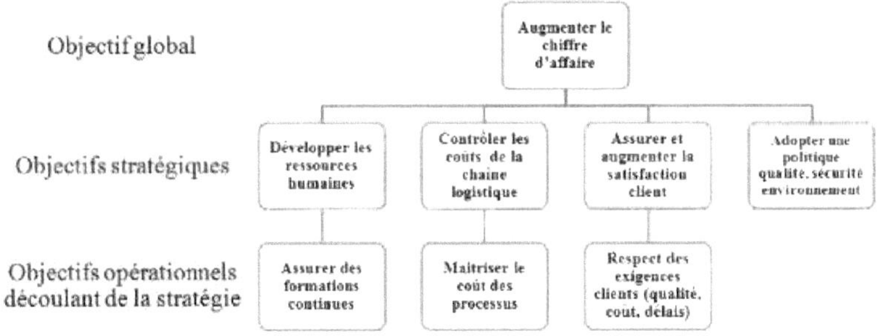

Figure 5.4 The translation of strategic objectives into operational objectives

5.4 Conclusion

In this chapter we have applied the first two phases of our methodology.

As the first phase was the description of the field of study, we presented the logistic chain of export of early fruits and vegetables, our perimeter of study which is a packing station and marketing of fruits and vegetables intended for export. Then, we presented the company by approaching its mission, its vision, its strategy and finish by its processes.

For the second phase, we analysed the upstream, internal and downstream logistics. The analysis of

81

the chain led to the identification of the processes constituting the logistics chain in order to arrive at its modelling and which allows the identification of the main processes that will be piloted by a dashboard. These processes are: supply, packaging and delivery and their interactions with the steering processes (planning and return).

Finally, a strategic diagnosis was carried out using SWOT analysis, which is a powerful tool for highlighting strategic orientations.

The next chapter will be devoted to the application of the third phase of our methodology which aims at building dashboards for the selected processes.

Chapter 6 Designing Supply Chain Process Dashboards

6.1 Introduction

The present chapter is devoted to the application of the third phase of our method of piloting a logistics chain by dashboard for the case of a fruit and vegetable packing station. This phase consists in designing operational dashboards by process of the chain (supply, packaging, delivery). We recall that our methodology is based on the following steps:

> Step 1: Process analysis ;

> Step 2: Risk identification and target setting;

> Step 3: Choice of indicators ;

> Step 4: Design a process dashboard.

6.2 Phase 3: Designing a process dashboard

Having modelled the supply chain in Chapter 5, and determined the main processes in our case study (Procurement, Packaging, Delivery), we now turn to the application of our approach to building a process dashboard.

6.2.1 Procurement process

6.2.1.1 Step 1: Supply process analysis

6.2.1.1.1 Identification of procurement processes

By means of identification cards, we will first identify, for the supply process :

- the title of the process ;

- the purpose of the process ;

- the person responsible for the process in question;

- customers externally and internally;

- the indicators monitored for this process.

Table 6.1 Procurement process identification sheet (Naciri et al, 2015c).

Title of the process	Procurement
Purpose of the process	Ensuring control of the supply of raw materials from the farms to the packing station Control of purchasing or any products or services that have an impact on quality Controlling all suppliers to the station through their evaluation.
Responsible for the process	Procurement Manager
Process client	Internal: packaging process External: customer, export group
Process indicators	Rate of deviation

6.2.1.1.2 The description

The procurement process has two main functions: the procurement of fruit and vegetables and the purchase of packaging and supplies.

- ♦♦ **Supply of fruit and vegetables**

The station manager compiles the export programme by variety from the group, reviews and approves it before passing it on to the packaging manager, the quality manager and the supply manager for implementation. If there are no anomalies, the supply manager communicates to each orchard the quantities and products to be picked.

The procurement process takes place as follows:

- Reception and arrival of fruit and vegetables

Reception is a key point in the fruit and vegetable supply process, it must be quick to avoid exposure of the goods to sunburn and ambient heat and must be careful to avoid crushing and injury to the fruit and vegetables.

A control is carried out at each reception, it consists in verifying the respect of the date before harvest by the producer and to carry out the approval of the goods.

- Identification of the raw material

This operation is carried out at the reception area, the products received are organised by batch and are identified by means of a label bearing the following information: Product, variety, producer, parcel, date of reception, time of reception, number of boxes, net weight of the batch.

- Storage

Warehousing is a very important operation, but it is not an obligatory step in the process of supplying fruit and vegetables. It allows on the one hand the adaptation and acclimatisation of the goods with the packaging conditions and on the other hand the organisation and management of the production flows.

- **Purchase of packaging and packaging supplies**

A weekly stock situation of packaging and packaging supplies and accessories is monitored. And the supply is triggered once the minimum safety stock is reached. These goods are subjected to a reception control to check the conformity with the preset prescriptions and to take the decision of acceptance or refusal of the goods.

- **Evaluation of suppliers and subcontractors**

The company ensures that the purchased product complies with the purchase requirements. These purchase specifications describe the product to be purchased and other requirements associated with it as well as possible. The company shall ensure the adequacy of these purchase specifications before communicating them to the supplier.

All suppliers and subcontractors with an impact on quality are evaluated. The evaluation of suppliers is carried out on the basis of their ability to meet the requirements of the order, including the quality management system requirements (Naciri et al, 2015c).

6.2.1.1.3 Process categorisation

According to the SCOR model, the procurement process corresponds to the Source process. To model this process, three types of supply are distinguished:

S1: Procurement for make-to-stock production, S2: Procurement for make-to-order production and S3: Procurement for design production.

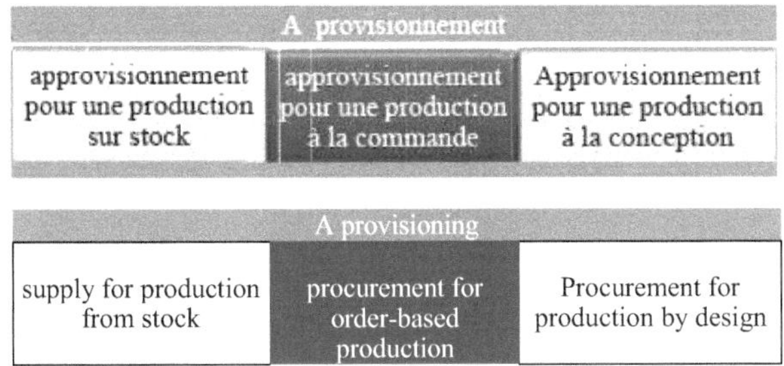

Figure 6.1 Categorisation of the Procurement process (Naciri et al, 2015c).

Our company's procurement method is S2: *procurement for make-to-order production.*

6.2.1.1.4 Establishing detailed process levels

According to the SCOR model, this process consists of five sub-processes:

ft S2.1 Plan procurement activities : Planning and managing product orders. The requirements for product orders are determined on the basis of the detailed supply plan.

52.2 Product acceptance: Product acceptance according to contractual requirements.

52.3 Product conformity check: To determine the conformity of the product to the requirements and criteria.

52.4 Product storage: The transfer of the accepted product to the appropriate storage location within the supply chain. This includes all activities related to packaging, transfer, and product and storage service or application.

52.5 Authorise Supplier Payment: Authorisation of payments from suppliers for products or services. This process includes collecting invoices and issuing cheques.

All the sub-processes of the SCOR model mentioned above are included in the process

and studied, except for the authorisation of payments to suppliers (Figure 6.2)

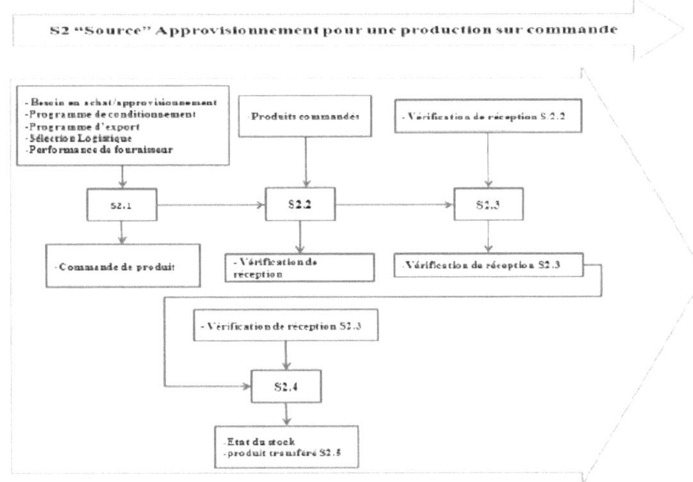

Figure 6.2 Level 3 process modelling of station supply (Naciri et al, 2015c).

6.2.1.2 Step 2: Risk identification and setting of operational objectives

6.2.1.2.1 Malfunctions / risks

The SCOR model allows for the simplification of business processes. We have compared the actual station process (Figure 6.4) with the standard procurement process of the model (Figure 6.3).

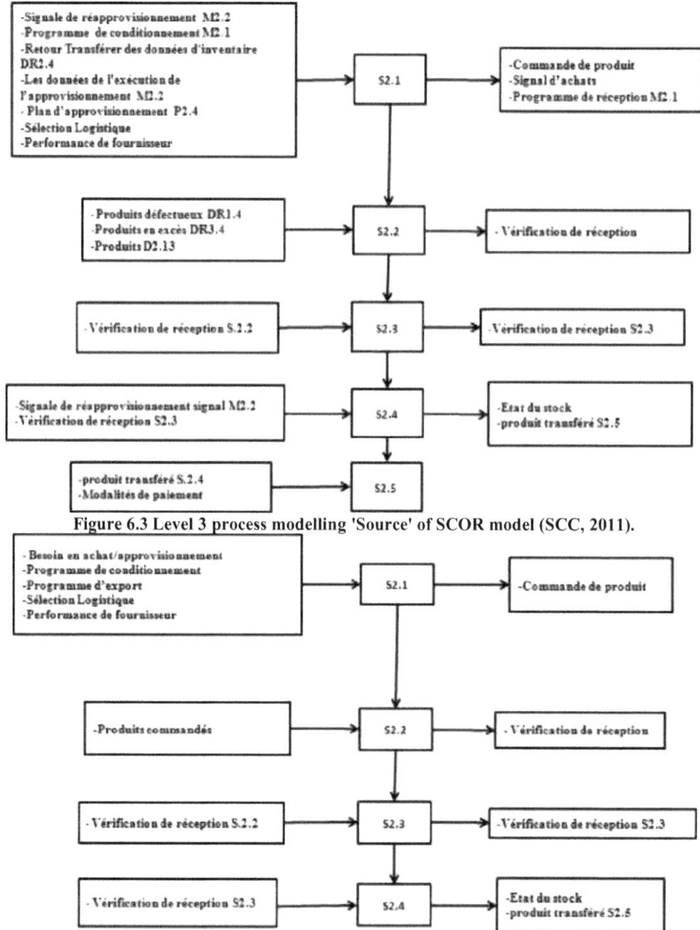

Figure 6.3 Level 3 process modelling 'Source' of SCOR model (SCC, 2011).

Figure 6.4 Level 3 process modelling of station supply (Naciri et al, 2015c).

Considering the process recommended by SCOR as "ideal", we were able to determine in this comparison how the procurement process differs from the SCOR process in terms of detecting dysfunctions. In fact, we distinguish three findings on the process during this comparison.

The procurement process proposed by the SCOR model, which consists of 5 sub-processes (from planning of procurement activities to supplier payment authorisation), is not considered by the station to be part of the procurement process: the sub-process "supplier payment authorisation" does exist on the station but is not located within the procurement process. In fact, the activity related to "supplier payment authorisation" takes place within the economic department. Therefore, there is a risk of authorising payment to suppliers before the conformity check, which risks accepting and receiving non-compliant products (Naciri et al, 2015c).

■ The "Replenishment signal" entry is not included in the "Procurement Activity Planning" activity, i.e. replenishment is not done by order of the packer;

■ The output "Receipt programme" is not located in the "planning of supply activities" activity. Therefore, the packaging manager (internal customer) is not informed of the availability of the raw material, which leads to a lack of anticipation and changes in the packaging schedule.

From these organisational dysfunctions, we have identified a number of

number of risks that have been validated by a working group (Table 6.2).

Table 6.2 Malfunctions and Risks associated with the procurement process (Naciri et al, 2015c).

Malfunctions	Associated risks
Replenishment is not done by order of packaging manager	Non-availability of raw material Delay in delivery
The packaging manager (internal customer) is not informed of the availability of the raw material	a lack of anticipation and changes in the packaging schedule.
Possibility of authorising payment to suppliers prior to the conformity check	accept and receive non-conforming products.

6.2.1.2.2 Determination of Operational Objectives

The operational objectives must be aligned with the strategy so that all teams drive the company towards performance. To determine the operational objectives for each process in the supply chain, we will integrate the risks from a process failure study (Figure 6.5).

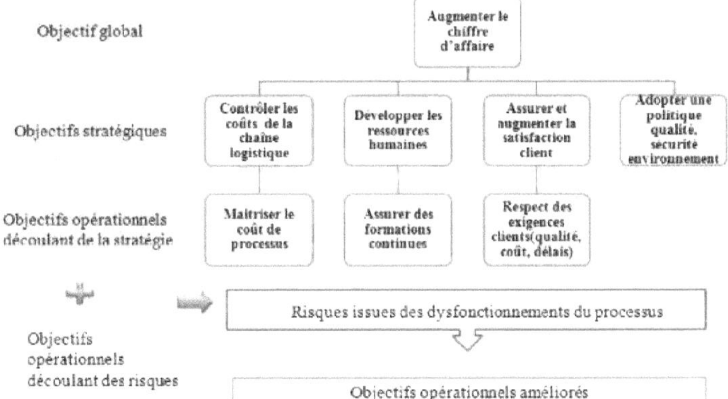

Figure 6.5 Deployments of objectives and introduction of risks (Naciri et al, 2014b).

Based on the risks identified in the previous step and the reflection of the working group, which led to the addition of other risks, we have associated one or more operational objectives. However, a number of objectives have been added which are derived from the company's strategy (Table 6.3).

Table 6.3 Risks and objectives of the procurement process (Naciri et al, 2015c).

Risks	Danger	Operational objectives
Delayed delivery	Manufacturing delay Non-compliance with export programme Degradation of product quality pending packaging	Have a safety stock Selection of suppliers
Product does not meet specifications and established	Product does not meet standards Product does not meet customer specifications Delayed delivery	Having a safety stock Selection of suppliers Inspection of the product on receipt
Non-availability of raw material	Delayed delivery	Have a safety stock Respect of delivery times for orders
Wrong Communication	Product does not meet customer specifications Delayed delivery Unaddressed claims	Selection of suppliers Compliance with customer requirements
High deviation in raw material	Non-compliance with the customer order	Reduce the gaps
-	-	Controlling costs supply
-	-	Providing continuous training

6.2.1.2.3 Identification of key process factors
We have broken down the operational objectives into key process factors, classifying them according to the Balanced Scorecard axes (Table 6.4)

Table 6.4 Objectives and Key Factors of the procurement process (Naciri et al, 2015c).

Axes	Operational objectives (OO)	Key process factors (CPF)
Financial	Controlling procurement process costs	Controlling the cost of supply services
Customer	Respect for delivery times	Reduce delivery delays
Process Procurement	Have a safety stock	Controlling stock levels Keeping a security stock
	Selection of suppliers	Purchase from agreed suppliers
		Evaluation of suppliers
	Reduce product deviations	Control on receipt of all products purchased
	Respect of delivery times for orders	Tracking orders
Organisational learning	Providing continuous training	Follow continuous training courses

6.2.1.3 Step 3: Selection of performance indicators

6.2.1.3.1 Filtering of indicators

Based on the SCOR model, we started by identifying level three indicators for the Procurement process.

We recall that this model gives general indicators by process. To this end, we propose to retain the balance of interactions between these indicators and those from the questionnaire published in our article (Naciri et al, 2015a) on important performance indicators in the industrial sector. The result of this work is given in Table 6.5.

Table 6.5 Interaction of SCOR indicators with procurement process indicators (Naciri et al, 2015c).
Common indicators
Increase in supplier payment times

Number of active suppliers monitored

Number of orders reviewed and in progress

Average time to process a purchase request

Service cost/ purchase turnover managed by the service

Service cost/savings generated by the service.

Average cost of placing an order

Average value of an order

Number of non-compliant batches / number of batches received

Average time to process a purchase request

Differences between quantities received and quantities ordered

Number of days of delay / number of late deliveries

6.2.1.3.2 Proposed indicators

From our study, we notice that the SCOR benchmark does not take into account the organisational and internal customer learning aspects. To this end, we find it necessary to propose indicators for these two axes. For this purpose, we have selected the indicators that are considered important for the industries in our study sample (Naciri et al, 2015a) and that are consistent with the company's objectives.

> ***For the financial axis :***

For the control of the cost of the supply process, we have proposed two indicators: Service Cost Rate and Average Cost of Placing an Order.

> ***For the internal process axis :***

For the different key process factors, we associate appropriate indicators:

> Controlling stock levels and maintaining safety stock: ***Stock coverage and stock levels;***

> Purchasing from agreed suppliers and monitoring of suppliers: ***Active suppliers monitored;***

> Control on receipt of all purchased products and reduce deviations: ***Compliance rate and deviation rate;***

> Tracking orders: ***Tracked orders ;***

> ***For the customer axis:***

To reduce delivery delays, we propose the following indicators: ***Delay rate.***

> ***For the organisational learning axis:***

For further training, we propose: the ***number of training hours*** as indicators.

The matrix below summarises the sources of our indicators (Figure 6.6).

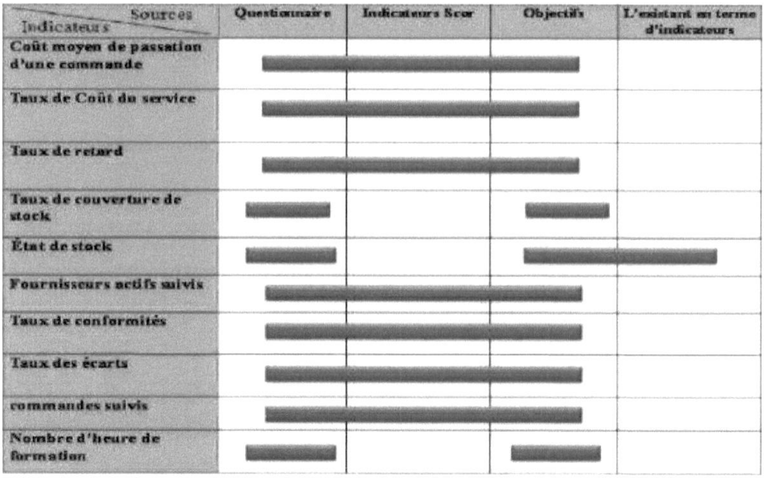

Figure 6.6 Indicators / Sources matrix (procurement process) (Naciri et al, 2015c).

6.2.1.4 Step 4: Drawing up a process dashboard

6.2.1.4.1 Dashboard design

In this final step, we summarise our work in the form of a table (Table 6.6) that represents the proposed performance indicators, which are classified according to the Balanced Scorecard axes. Thus we present the method of calculation of these indicators, the target performance, the frequency and the person responsible for collecting the data.

PI: Process Indicators. MC: Calculation mode. P: Performance. F: Frequency.

R: Responsible person. AP: Action Plan. T: Quarterly. Rg: Regular. app: Procurement

Table 6.6 Procurement process dashboard (Naciri et al, 2015c).

AXES	IP	MC	P	F	R	PA
Financial	Average cost of placing an order	Average cost of procurement of a order		T	R app	Identify sources of cost increase
	Service cost rate	Service cost/ Purchase turnover managed by the service	<2%	T	R app	Verify and control the service cost. Optimal supply management
Internal client	Delay rate	Number of days overdue / number of late deliveries	<10%	T	R app	Compliance with deadlines
Process Procurement	Rate stock coverage	average consumption for the period/average stock for the period	--	Rg	R. shop	Ensuring the availability of raw materials
	Stock status	Stock status	--	Rg	R app	Replenishment
	Active suppliers monitored	No. of active suppliers monitored	--	T	R app	Tracking suppliers
	Rate conformities	No. of non-compliant batches / No. of batches received	<5%	Rg	R app	Identification of the source of non-compliance and deviations
	Rates for deviations	Deviations quantities received / quantities ordered	<5%	Rg	R app	
	- follow-up orders	- Number of orders received and in progress	--	Rg	R app	Tracking orders

Organisational learning	Number of hours of training	Number of hours of training	--	T	R quality	Follow continuous training

6.2.1.4.2 Causal link between indicators

After proposing the process steering indicators, the causal relationship between indicators must be visible on the four axes of the dashboard. We represent

Figure 6.7 shows the causal link between the procurement process indicators.

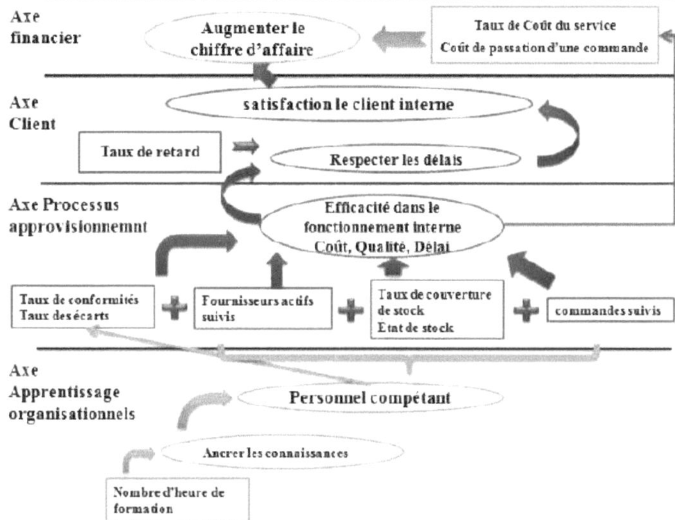

Figure 6.7 Causal linkage of procurement process indicators

6.2.2 Packaging process

6.2.2.1 Step 1: Packaging process analysis

6.2.2.1.1 Identification of packaging processes

In the same way as the identification of supply processes, we determine for the packaging process :

- The title of the process ;

- The purpose of the process ;

- The process owner in question ;

- External and internal customers;

- The indicators monitored for this process.

Table 6.7 Process identification sheet Packaging (Naciri et al, 2015b).

Title of the process	Packaging
Purpose of the process	the execution of the packaging programmes while ensuring good product quality, respecting the deadlines set beforehand and optimising performance.
Responsible for the process	Packaging Manager
Process client	Internal: delivery process External: final customer, export group
Process indicators	Number of customer complaints

6.2.2.1.2 The description

The packaging of fruits and vegetables is a variable process depending on the product and the customer's requirements. We choose to deal with the case of tomato as it represents 90% of the total production (Naciri et al, 2015b).

♦♦♦ Tomato packaging:

Pouring : The pouring consists in pouring the tomato in the packaging line, it is done manually. During the pouring process the tomato is subject to the danger of injury and crushing. For this reason, pouring should be done with care and attention.

Washing and showering: The purpose of washing and showering is to remove dust, dirt and other impurities from the tomato fruit. The fruits are washed with treated and controlled drinking water.

Dewatering: Dewatering is the removal of water droplets from the surface of the fruit. This operation is carried out mechanically for round tomatoes and manually for grape tomatoes and cherry tomatoes.

Drying: This consists of removing excess moisture from the surface and inside the skin of the fruit. The fruit is passed under hot ventilated air at a temperature of 40 to 60°C.

Sorting: The purpose of sorting is to eliminate sorting errors.

Grading and classification: Classification of fruits according to their diameter.

Packaging: After selection and grading, the fruit is packed in cardboard boxes. The box varies according to the product.

Labelling: Labelling is a very important operation. It allows to trace the production, the recall of the product and to inform the consumer about the product and its origin.

Palletising: Palletising is an operation which consists of collecting and placing packages on a pallet for export. It must be done properly and with great care in order to avoid the packages falling and consequently avoiding the crushing of the product.

Storage: The final products ready for shipment are stored in the cold room awaiting shipment. The tomatoes are stored in the refrigerators at a temperature of 8 to 10°C and a relative humidity of 95%.

Treatment of off-cuts: Off-cuts from the packaging process are destined for the local market. Prior to marketing, the off-cuts are sorted and graded and sold on the local market.

Non-conformities: All non-conformities detected during production operations are dealt with in accordance with the procedure.

6.2.2.1.3 Process categorisation
According to the SCOR model, the conditioning process corresponds to the MAKE. To model this process, three types of production are distinguished: M1 make-to-stock, M2 make-to-order and M3 design-to-order.

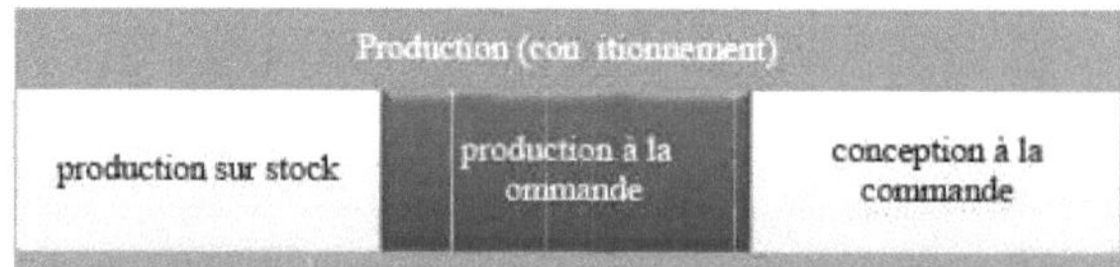

Figure 6.8 Categorisation of the Make process (Naciri et al, 2015b).

The ordering production of our company is M2: production (packaging) at the mode.

6.2.2.1.4 Establishing process detail levels
According to the SCOR model this process consists of seven sub-processes:
- M2.1: Plan production activities ;

- M2.2: Raw material preparation ;
- M2.3: Production and testing;
- M2.4: Packaging and wrapping ;
- M2.5: E ape of finished product;
- M2.6: Authorise delivery of finished product;
- M2.7: Scrap disposal.

The first module M2.1 aims at programming the production activities, i.e. the production schedule, the raw material according to the planning and the demand for the necessary means. In our case, this module is not integrated in the packaging process.

The next module M2.2 (raw material preparation) is necessary to properly place the raw materials from a storage area to a specific point of use. For this reason, the module has three main tasks: It must be possible to check whether there is raw material available and if so, send a signal to the product warehouse to release the raw material to production. Finally, the raw material and production plans must be updated.

For modules M2.3, M2.4, they will be combined into one module (M2.3 packaging and packing). This M2.3 module aims to ensure a better packaging process according to the product and customer requirements.

For the two modules M2.5 and M2.6, they will be combined into a single module (M2.4 finished product stage and authorise delivery). The purpose of module M2.4 is to transport the packaged products to a finished product storage location.

Module M2.7: Scrap disposal, which aims at the disposal of all production waste. In our case study, this module is equivalent to process M2.5 (Disposal of waste from packaging processes) (Figure 6.9).

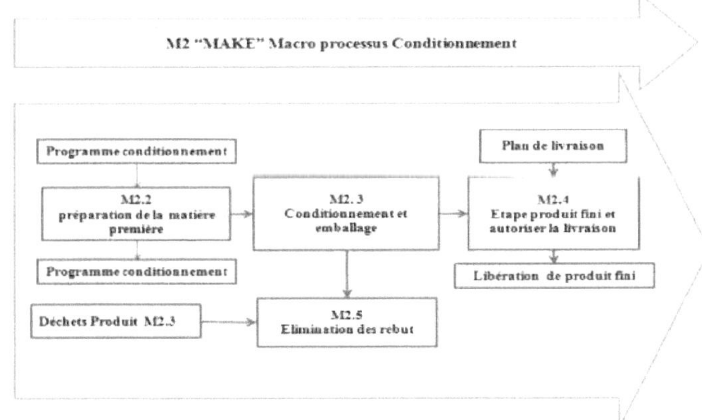

Figure 6.9 Level 3 modelling of packaging process (Naciri et al, 2015b).

6.2.2.2 Step 2: Risk identification and setting of operational objectives

6.2.2.2.1 Malfunctions / risks

In the same way as for the supply process, we compared the actual station process (Figure 6.11) with the standard make process of the model (Figure 6.10) to detect malfunctions.

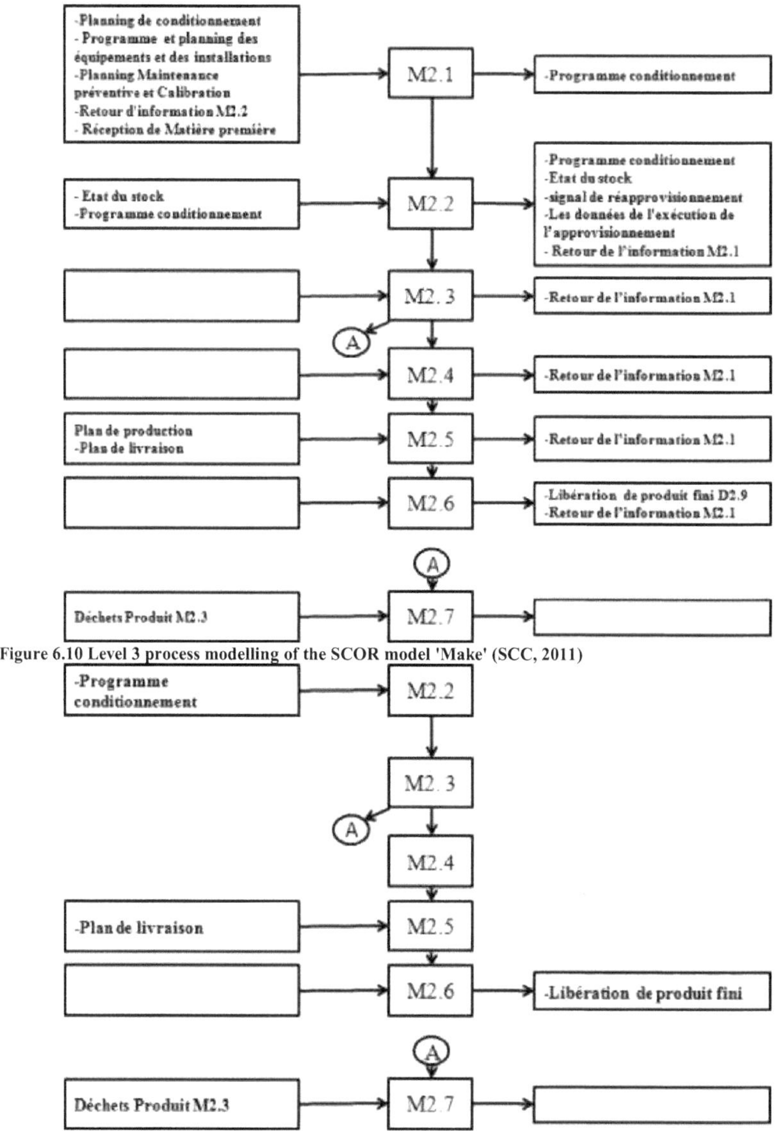

Figure 6.10 Level 3 process modelling of the SCOR model 'Make' (SCC, 2011)

Figure 6.11 Level 3 process modelling of station conditioning (Naciri et al, 2015b).

When comparing the station's conditioning process with that of SCOR, two remarks can be made:

■ The activity related to the M2.1 SCOR sub-process "planning of packaging activities" exists in the station but is not located within the packaging process. It is located within the planning process and the station manager carrying out this planning is therefore far from the concerns of the packaging process. A number of risks can be identified. For example, the execution of the packaging programme by the station manager without much knowledge of the operational constraints can lead to errors in the planning (overlapping of the production programme with the preventive maintenance planning). On the other hand, the packaging manager has a clear and efficient view of the process to establish a packaging programme.

■ The activity M2.2 "preparation of raw material" is only concerned with the reception and storage of raw material, while the SCOR activity includes other tasks (stock status, replenishment signal, supply execution data, feedback to M2.1).

From these organisational dysfunctions, we have identified a number of risks:

o Stoppage of production due to unavailability of raw material;

o Non-compliance with the packaging programme;

o Delayed delivery - no customer satisfaction;

o Degradation of the quality of the raw material.

6.2.2.2.2 Risks/ objectives

Similarly for the procurement process risk identifications. Based on the risks identified in the previous step and other operational risks, we link risks to objectives.

Table 6.8 Risks/objectives packaging process (Naciri et al, 2015b).

Risks	Operational objectives
Non-compliant product	Controlling product quality Provide ongoing product training
Failure to meet customer requirements	Reduce customer complaints
Delayed delivery	Adhere to the conditioning programme
Unforeseen machine failures	Increase machine availability.
Non-availability of raw material	Controlling stock levels
Non-compliance with the programme	Adhere to the conditioning programme
--	Controlling packaging costs

6.2.2.2.3 Identification of key factors in the packaging process
From the operational objectives, we determine the key process factors (Table 6.9).

Table 6.9 Objectives / CFP conditioning process (Naciri et al, 2015b).

Axis	Operational objectives (OO)	Key process factors (CPF)
Financial	Controlling costs	Controlling the costs of non-quality
	Increase annual production	Increase in tomato production
Customer	Reduce customer complaints	Respecting the delivery time Meeting customer quality requirements
Packaging process	Adhere to the conditioning programme	Adhere to the conditioning programme
	Increase the availability machine.	Application of the preventive maintenance programme Implementation of the infrastructure maintenance programme
	Controlling product quality	Reduce product deviations
	Controlling the stock situation	Keeping a security stock
Organisational learning	Provide ongoing product training	Compliance with the training plan

6.2.2.3 Step 3: Selection of performance indicators

6.2.2.3.1 Filtering of indicators

We have retained the interaction balance between the production indicators generated by the SCOR model and our bank of indicators. The result of this work is given in Table 6.10

Table 6.10 Common indicators of the conditioning process (Naciri et al, 2015b).

Common indicators

ASR = Availability rate x Performance rate x Quality rate

Number of late finished Production Orders ^ total number of Production Orders
Time to return to production ^ number of pieces returned to production
Actual production rate ^ target production rate
Standard cycle time / Actual cycle time
Actual cycle time/ Ideal cycle time
Number of defects produced due to raw material quality ^ total number of defects Monthly energy consumption.
Amount of waste recycled
Cost of damaged products due to staff errors ^ total cost of damaged products
Cost of product defects due to raw material quality ^ Total cost of defects
Total production cost ^ total number of units produced
Production cost ^ Cost of sales
Costs associated with machine downtime
Production cost vs last year vs budget

6.2.2.3.2 Proposed indicators

> *For the financial axis:*

We have kept the performance indicator used by the company that responds to the increase in tomato production: Increase in tomato production.

For the control of the cost of non-quality, we proposed two indicators: *% damaged products due to staff errors and % product defects due to PM quality.*

> *For the internal process axis :*

For the packaging process, compliance with the delivery date depends on compliance with the packaging programme. Therefore, we have proposed an indicator that responds to these two FCPs: *Production Order Achieved.*

With regard to the implementation of the preventive maintenance programme and the reduction of product deviations, we have assigned the indicator: *Synthetic efficiency rate (SER).*

For the safety stock, the packaging manager must monitor the stock situation at all times and have a minimum stock to prevent shortage. For this factor we propose the supply process indicator: *Stock status.*

> *For the customer axis:*

Compliance with customer quality requirements is reflected in the indicator: *Number of customer complaints.*

> *For the organisational learning axis:*
We select the following indicators: *Rate of production stoppages due to lack of staff training.*
We propose to summarise the sources of our indicators in the form of a matrix (Figure 6.12) that links the indicators with the sources from which they were obtained.

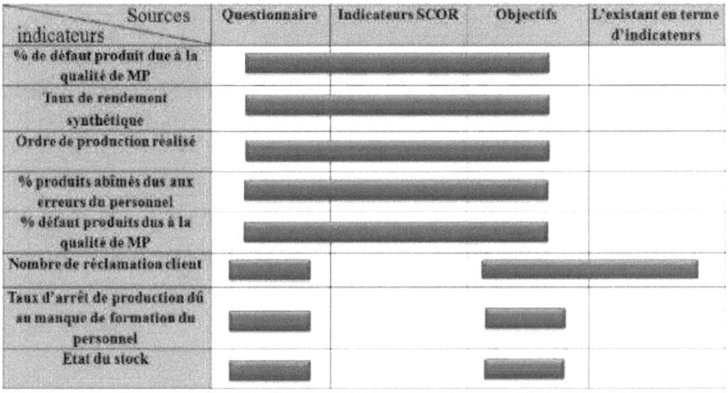

Sources / indicateurs	Questionnaire	Indicateurs SCOR	Objectifs	L'existant en terme d'indicateurs
% de défaut produit due à la qualité de MP				
Taux de rendement synthétique				
Ordre de production réalisé				
% produits abîmés dus aux erreurs du personnel				
% défaut produits dus à la qualité de MP				
Nombre de réclamation client				
Taux d'arrêt de production dû au manque de formation du personnel				
Etat du stock				

Figure 6.12 Indicators / Sources matrix (conditioning process) (Naciri et al, 2015b).

6.2.2.4 Step 4: Development of a process dashboard

6.2.2.4.1 Packaging dashboard

Finally, we summarise our work in a table (Table 6.11) that represents the proposed performance indicators, which are classified according to the Balanced Scorecard axes. Thus we present the method of calculation of these indicators, the target performance, the frequency and the data collector.

PI: Process Indicators. MC: Calculation mode. P: Performance. F: Frequency.

R: Responsible person. AP: Action Plan. T: Quarterly. Rg: Regular. Cd: Packaging

Table 6.11 System of packaging process performance indicators (Naciri et al, 2015b).

AXES	IP	MC	P	F	R	PA
Financial	% products damage due to the errors of the staff	Cost of damaged products due to staff errors ^ total cost of damaged products	< 1%	T	R quality	Formation of staff
	% defect products due to the quality of PM	Cost of product defects due to raw material quality ^ Total cost of defects	< 1%	T	R quality	Evaluation of Suppliers Review Reception Control
Internal client	Number of customer complaint	Number of customer complaints	<3	T	RCd	Respect customer requirements
Packaging process	% of product defects due to PM quality	Number of defects produced due to raw material quality ^ total number of defects	< 1%	T	R quality	Evaluation of Suppliers Review Reception Control
	Rate synthetic performance	Availability rate x Performance rate x Quality rate	> 70%	T	RCd	Application of the preventive maintenance programme Reducing product deviations
	Order from actual production	Number of late finished Production Orders ^ total number of Production Orders	> 75%	T	RCd	Eliminate malfunctions due to non-compliance with the packaging programme
	Stock status	Current stock/storage capacity	>10%	Rg	R shop	Replenishment
Organisational learning	Rate of production stoppage due to lack of staff training	Production stop due to lack of staff training / total production stoppages	>90%	T	R quality	Compliance with the training plan

6.2.2.4 .2 Causal link between indicators

Figure 6.13 shows the causal link between the conditioning process indicators.

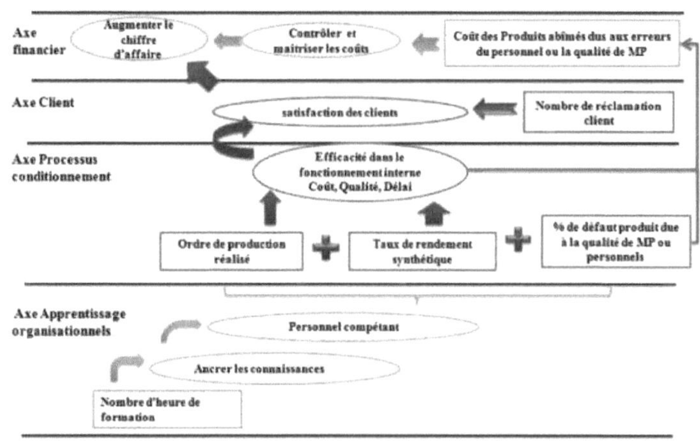

Figure 6.13 Causal linkage of packaging process indicators

6.2.3 Delivery process

6.2.3.1 Step 1: Process analysis

4.2.3.1.1 Process identification

The table (Table 6.12) shows the title of the delivery process, the purpose of the process, the person responsible for the process in question, *the* external customers and the indicators monitored for this process.

Table 6.12 Delivery process identification sheet

Title of the process	Delivery
Purpose of the process	Ensuring the smooth shipment of finished products to the exporting group/end customer
Responsible for the process	Delivery Manager
Process client	External: customer, export group
Process indicators	-

6.2.3.1.2 The description

The delivery process of the station is as follows:

J Depending on the expected capacity and the orders held, a delivery date is committed and planned;

J Orders are analysed to determine the groups that result in the lowest cost/best service and transport performance so that modes of transport are selected and loads are constructed;

J Verification, receipt of product and determination of storage location for goods received either from manufacture or source;

J Specific carriers are selected by lowest cost per trip;

J Product loading and generation of shipping documents ;

J Shipment of finished products to the exporting group or directly to the end customer in the case of a specific order;

J Invoicing: the invoicing and payment process begins.

6.2.3.1.3 Categorisation of the delivery process

The Deliver process is concerned with the dispatch of products and services to customers to meet expected, planned or actual demand. The process categories are as follows:

- D1 :   Deliver Stocked Product   ;
- D2 :   deliveryofproducts   its to order (Make-to-Order);
- D3 :   Delivery of designed   projects to order (Engineer-to-Order);
- D4 :   Retail distribution   (Deliver Retail roduct).

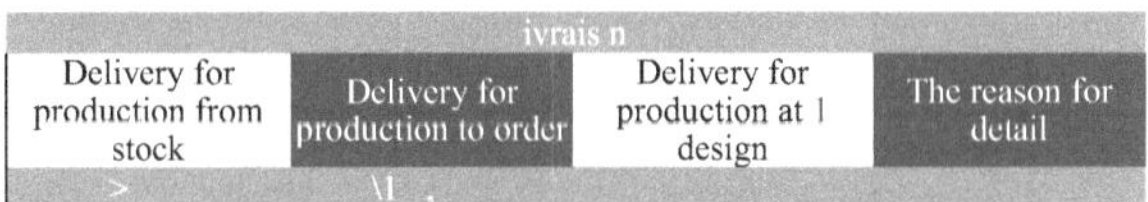

Figure 6.14 Categorisation of the delivery process

The delivery method of our company is D2: ***delivery for production to order,*** and D4: ***retail delivery.***

The delivery mode D4 "retail delivery" is for items that are destined for the local market, so in the following we will only consider the delivery mode D2 "*delivery for production to order*".

6.2.3.1.4 Establishing process detail levels

According to the SCOR model this process consists of 15 sub-processes:

D.2.1 Receiving orders: Receiving orders and responding to customer enquiries.

ft D2.2 Order processing: Receive orders from customers and enter them into the company's order processing system. Configure the product to customer specifications, based on standard available parts or options.

99

D2.3 Allocation: Identify the expected capacity for specific orders, and assign a delivery date.

D2.4 Consolidate orders: Analyse orders to determine the groups that result in the lowest cost/best service and transport performance.

D2.5 Construct loads: Modes of transport are selected and effective loads are constructed.

D2.6 Road deliveries: loads are grouped and routed by mode, route, and location

D2.7 Carrier selection: Specific carriers are selected by lowest cost per trip and shipments are classified and discounted.

D.2.8 Receiving the product: Activities such as checking, receiving product, determining storage location and recording location for goods received either from manufacture or source. May include quality inspection.

D.2.9 Product selection : Checking stock availability, selection of transport vehicle, product selection and delivery of product to the packing area.

D.2.10 Packing: Activities such as sorting of products, packing, barcode labelling, etc., are carried out by the company. And the delivery of products to the shipping area for loading.

D.2.11 Load Product and Generate Shipping Documents: The series of tasks, including product, modes of transport, and generation of necessary documentation, customer, carrier and government requirements. Shipping documentation includes the invoice. Possibly check customer credit.

D.2.12 Shipping: The process of shipping the product to the customer's site.

D.2.13 Receiving and checking the product by the customer : The receiving process verifies that the order has been shipped complete and that the product meets the delivery requirements.

ft D.2.14 Product installation : Where necessary, the process of preparing, testing and installing the product at the customer's site. The product is fully functional at the end.

D.2.15 Invoice: A signal is sent to the financial organisation that the order has been dispatched and that the invoicing and payment process should start. Payment is received by the customer in the payment terms of the invoice.

To simplify, we will combine the two modules *D.2.1 Order reception* and *D2.2 Order processing* into one module D2.1 *Order reception and Processing*. The modules D2.4-D2.7 will be renamed *"Dispatch"*. Modules D2.8-D2.10 will be renamed *"Delivery Preparation + Packaging"*. Modules D2.11 and D2.12 will be *renamed "Shipping"*. Figure 6.15 models the level three delivery process of the station.

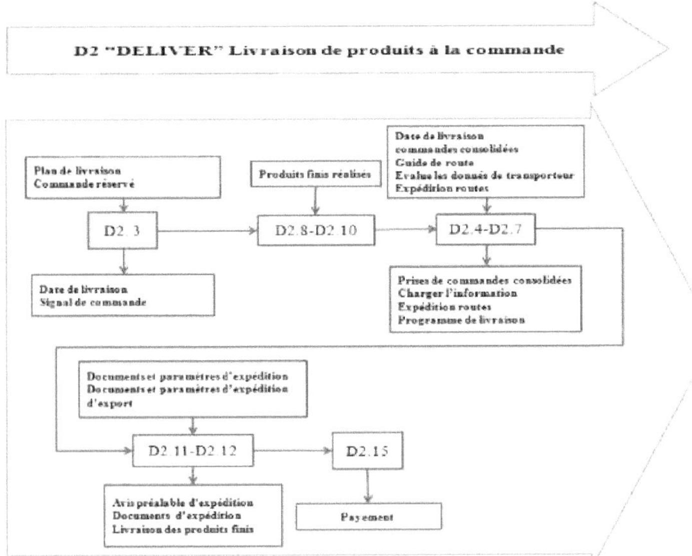

Figure 6.15 Level 3 delivery process modelling

6.2.3.2. Step 2: Risk identification and setting of operational objectives

6.2.3.2.1 Malfunctions / risks

We compared the station's actual process to the standard delivery process of the SCOR model (Figure 6.16). In this comparison we were able to determine how the station's process (Figure 6.17) differs from the standard SCOR process.

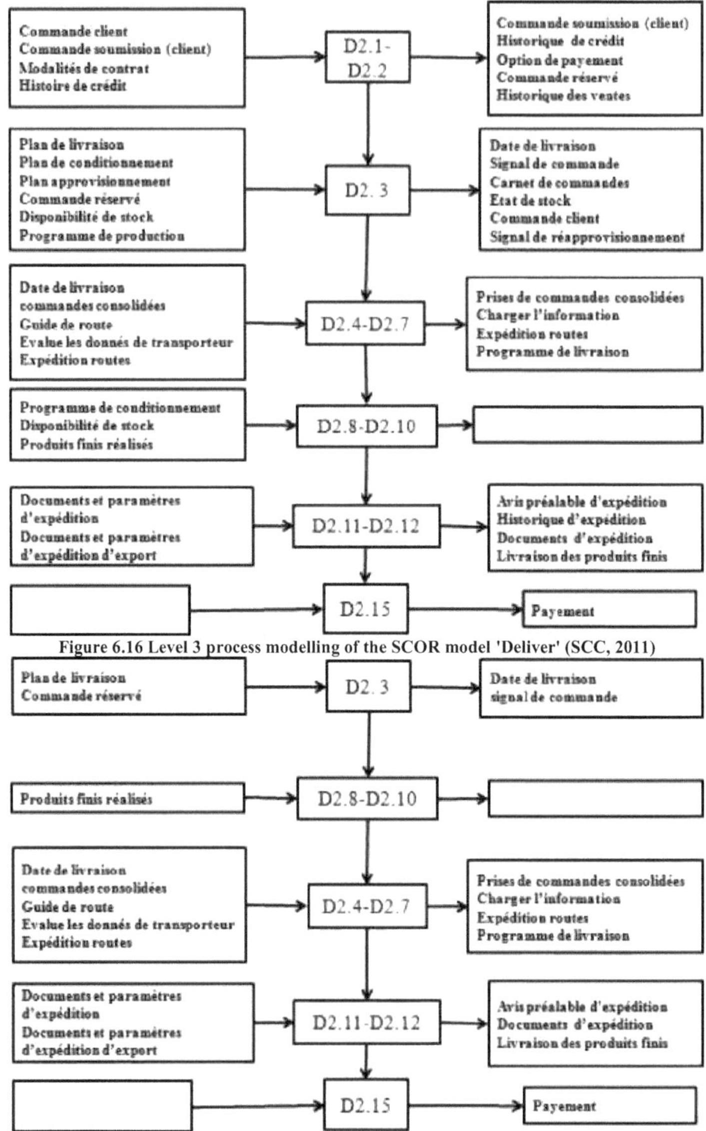

Figure 6.16 Level 3 process modelling of the SCOR model 'Deliver' (SCC, 2011)

Figure 6.17 Level 3 process modelling of station delivery

The delivery process proposed by the SCOR model, consisting of 15 sub-processes (from order processing to payment of the delivery carrier), differs from the station process in the following way:

■   The activities related to the SCOR sub-processes D2.1 and D2.2 "reception and processing of orders" are not integrated in the station's delivery process. Indeed, they exist on the station but are not part of the delivery process. They take place in the commercial department.

From this comparison, we detect that the commercial staff is far away from logistical concerns. A number of risks can be identified. For example, the placing of customer orders by sales assistants without much knowledge of transport issues can lead to some errors in these orders.

■ The activities related to the sub-processes D2.4 to D2.10 "dispatch and delivery preparation" are in the delivery process but in a different sequence than recommended by SCOR. The product of the station, after being assigned to an order, is first prepared and then stored in the cold rooms. Then the logistics department intervenes in the process with the shipment management to order the transport of the product. However, SCOR's delivery process recommends the opposite. Indeed, the flow proposed by SCOR seems to be faster than the one of the station because if the transport is ordered before or while the product is in physical preparation, time is saved on the delivery process (preparation in hidden time).

■ The sub-process "reception and installation of the product on the customer site" (D2.13 and D2.14) does not exist in the station delivery process. This is logical, as it is a sub-process that does not concern the company's activity.

Based on these malfunctions, we group together the risks associated with them in a table (Table 6.13).

Table 6.13 Delivery Process Malfunctions and Risks

Malfunctions	Associated risks
Orders are processed by the sales department	Poor delivery process management Poor communication Delayed delivery
Lack of training on the delivery process	Order processing error
Waste of time due to the sequence of activities	Increasing the delivery preparation time

6.2.3.2.2 Risks/ objectives

At this level, we have drawn up a list of risks that affect the proper functioning of the process. Thus, we have associated one or more operational objectives with each risk.

Table 6.14 Delivery process risks / objectives

Risks	Danger	Operational objectives
Delayed delivery	Non-compliance with export programme	Respecting delivery times
Order processing error	Product does not meet customer specifications Delayed delivery	Provide continuous training Respecting order processing times
Poor process management Delivery	Delayed delivery Unprocessed claims Non-compliance with the customer order	Respecting delivery times
Increasing the delivery preparation time	Delayed delivery	Controlling and respecting delivery preparation times
--	--	Controlling delivery costs
--	--	Providing continuous training

6.2.3.2.3 Identification of key process factors

We have broken down the operational objectives into key process factors, classifying them according to the Balanced Scorecard axes (Table 6.15)

Table 6.15 Objectives and FCPs delivery process

Axis	Operational objectives (OO)	Key process factors (CPF)

Financial	Controlling costs	Controlling delivery process costs
Customer	Customer satisfaction	Reduce customer complaints Reduce delivery delays
Delivery process	Respecting delivery times	Adhere to the delivery schedule Respecting the export schedule
	Respecting order processing times	Reduce order processing time
	Controlling and respecting delivery preparation times	Reduce delivery preparation time
Organisational learning	Providing continuous training	Follow continuous training

6.2.3.3 Step 3: Selection of performance indicators

6.2.3.3.1 Filtering of indicators

The common indicators between the delivery process indicators of the SCOR model and the result of our questionnaire published in our paper (Naciri et al, 2015a) are grouped in Table 6.16.

Table 6.16 Common indicators of the delivery process
Common indicators

The time associated with receiving, entering and validating a customer order.
Time of execution of orders,
of orders delivered in full, average time associated with the shipment of products Delivery variances
Number of deliveries per hour ^ Total number of deliveries during the same period
Capacity used (m3) ^ capacity available (m3) during the same period
Annual number of deliveries (or tons, volumes of books...)
Shipping cost Product
Transport cost ^ Cost of sales
Total transport cost

6.2.3.3.2 Proposed indicators

> *For the financial axis:*

For the control of the delivery process cost, we have proposed two indicators: ***Product shipping cost*** and ***Total transport cost.***

> *For the internal process axis :*

To reduce order processing time, we need to monitor the following indicators: ***Time to complete orders*** and ***Time associated with receiving, entering and validating a customer order.***

In order to reduce the preparation time for delivery, we monitor 1 indicator: ***The average time associated with the shipment of products.***

For compliance with the export schedule and the delivery programme, we propose the indicator: ***Delivery variances.***

> *For the customer axis:*

To reduce customer complaints and delivery delays, we propose these indicators: ***Number of customer complaints*** and ***Delay rate.***

> *For the organisational learning axis:*

In order to follow the continuous training, we suggest to follow; The number of training hours.

A summary of the sources of our performance indicators is given in the form of a matrix (Figure 6.18).

Sources Indicators	Questionnaire	SCOR indicators.	Objectives	L' existing in terms of indicators

Expedition cost Product				
Total transport cost				
Delay rate				
Number of claims				
Order execution time.				
The time it takes to receive, enter and validate an order				
The average time associated a Γ exp edition of the products				
Delivery variances				
Number of hours of training				

Figure 6.18 Indicators / Sources matrix (delivery process)

6.2.3.4 Step 4: Development of a process dashboard

6.2.3.4.1 Delivery dashboard design

Finally, we summarise our work in the form of a table (Table 6.17) that represents the proposed performance indicators, which are classified according to the Balanced Scorecard axes. Thus we present the method of calculation of these indicators, the target performance, the frequency and the data collector.

PI: Process Indicators. MC: Calculation mode. P: Performance. F: Frequency.

R: Responsible person. AP: Action Plan. T: Quarterly. Rg: Regular. Liv: Delivery

Table 6.17 Delivery process dashboard

AXES	IP	MC	P	F	R	PA
Financial	Shipping cost Product	Cost shipping cost Product	-	T	R Liv	Identification of sources of waste
	Total transport cost	Total transport cost	-	T	R Liv	Reconfiguring Process of delivery
Customer	Delay rate	Cumulative number of days late / number of late deliveries	<10%	T	R Liv	Respecting the delivery time
	Number of customer complaint	Number of customer complaints	<3	T	R Liv	Respect customer requirements
Process Delivery	Time the execution of orders.	Order execution time	-	Rg	R Liv	Decrease the time of execution and processing of orders
	The time associated with receiving, entering and validation of an order from the customer	The time associated with receiving, entering and validating of a customer's order	-	R	R Liv	Reduce the time associated with receiving, entering and validating an order
	The average time associateda the expedition of products	The average time associated with the shipment of products	-	Rg	R Liv	Reconfiguring the process delivery
	Delivery variances	Valuation of differences / total value of products transported during the same period	<5%	T	R Liv	Determining the sources of discrepancies
Organisational learning	Number of hours of training	Number of hours of training	-	T	R Liv	Follow continuing education

4.2.3.4.2 Causal link between indicators

Figure 6.19 shows the causal link between the delivery process indicators.

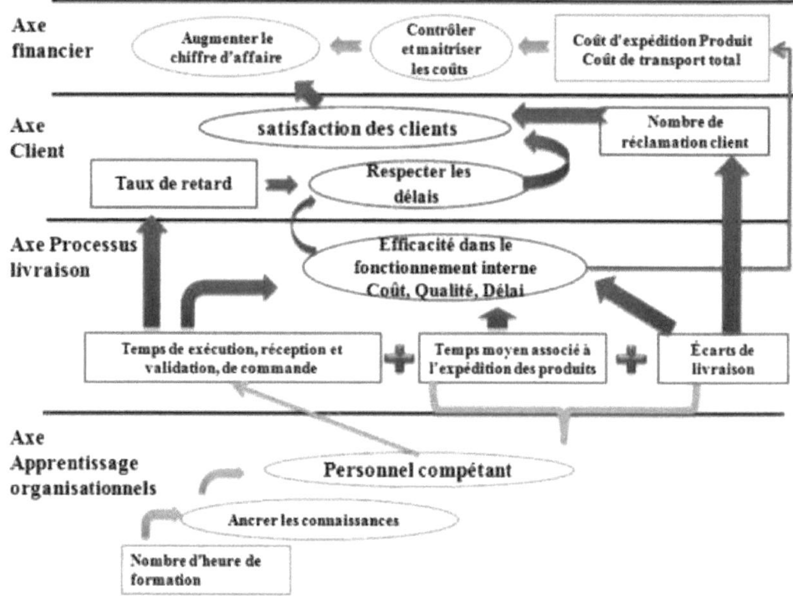

Figure 6.19 Causal linkage of delivery process indicators

6.5 Conclusion

At the end of this chapter, we have developed three operational dashboards for the management of the main processes of the station's supply chain.

In order to achieve our objective, we started with the process analysis (identification, description and modelling), which is an important step, as it allows us to understand the functioning of the process within the supply chain.

Then, the diagnosis of the processes by comparing the current process with the SCOR model process, led us to highlight the dysfunctions and risks incurred.

As performance indicators drive business activities or processes in order to avoid risks and align strategy and operations, we have identified operational objectives by process and those associated with the risks identified.

Inspired by the Balanced Scorecard approach in the breakdown of the chosen operational objectives into process key factors and then into indicators, we were able to propose our own indicators by process and by performance axis.

The choice of indicators responding to the key process factors requires a reflection on the relevance of the indicators. For this purpose, we used the indicators proposed by the SCOR model and the indicators selected by the companies interviewed by our questionnaire. Indeed, the intersection between these two sources constituted our database of indicators. The interest of this step lies in the adoption of indicators that are adapted to the industrial sector and approved by the world reference system for supply chain management, the SCOR model.

In order to select relevant indicators that really contribute to decision making, we have assigned for each key process factor a performance indicator with reference to our database.

Finally, we have summarised our work in the form of a dashboard that represents the proposed performance indicators and classified them according to the axes of the Balanced Scorecard. Thus we have presented the method of calculation of these indicators, the target performance, the frequency and the person responsible for collecting the data.

Chapter 7: Designing a Global Supply Chain Scorecard

7.1 Introduction

In this last chapter, we will carry out a recent and detailed bibliographical study of the main global indicators of a supply chain in order to come out with a synthesis of the global indicators of the supply chain. The last part will be devoted to the application of phase 4 of our proposed approach, the aim of which is to build a global dashboard of the supply chain for our case study.

7.2 Application of Phase 4: Designing a Supply Chain Scorecard

7.2.1 Determination of the performance axes.

To design our TDB, we start from the strategic objectives of the company determined in phase 2 of our methodology. We recall that these objectives are :

J Controlling supply chain costs;

J Ensure customer satisfaction;

J Develop human resources to build a team recognised for its performance in the areas of activity.

J adopt a quality, safety and environmental policy for the packaging and marketing of fruit and vegetables for all its activities and sites, based on the BRC, HACCP and Sustainable Development standards. And this to maintain a high level of quality in order to retain an increasingly demanding clientele and to face up to very aggressive competition.

We note that the company aims to adopt a quality, safety and environmental policy that aims at continuous customer satisfaction:

Quality, in order to :

- To provide a healthy product that complies with all applicable quality standards and regulations at the lowest possible cost;

- To meet customer requirements in terms of deadlines and technical specifications;

- Continuously improve the quality system by optimising and enhancing its human and material resources.

- Security, in order to :

- To ensure the safety of personnel within the company against any danger;

- Ensuring professional health and safety conditions for all staff.

- Environment, in order to :

- Protecting the environment by adopting techniques and practices that conserve wildlife.

Done, it seems necessary to add to our scorecard, which takes into account the financial (costs) and non-financial (quality, deadlines, organisational learning) dimensions, indicators capable of measuring social and societal issues in another axis. This observation is validated by Hockerts (2001), who proposed the elaboration of a Sustainability Balanced Scorecard (SBSC), an extension of the initial TBP but partly composed of indicators measuring the environmental and social performance of companies.

Supizet (2002) has suggested, alongside the SBSC, the concept of the Total Balanced Scorecard (TBSC) whose model is based on a series of six causal relationships between stakeholders: shareholders, customers, users, the company as a legal entity, partners, employees and the community. In 2012, Maria (2012) applied the HWC to implement effective strategies in which economic, social and environmental aspects were integrated into an integrated system for assessing sustainable performance. Tsalis et al (2013) proposed the HWC as an essential strategic management tool to increase awareness of corporate responsibilities.

The SBSC structure has been proposed to reflect the overall performance on the basis of the BSC, integrating sustainability related parameters. Furthermore, the SBSC not only detects the strategic environmental and social objectives of a company, but also improves the transparency of value-added potentials that emerge from social and ecological aspects (Zhao, 2015).

7.2.2 Reflection

Our thinking for the design of the supply chain dashboard is as follows:

We start with the operational objectives of the processes, which are selected from the strategic objectives, and the analysis of each process in the chain. We propose to review and validate these in order to take into account the expectations and ambitions of the field actors, and the operational constraints in formalising the strategy (Figure 7.1).

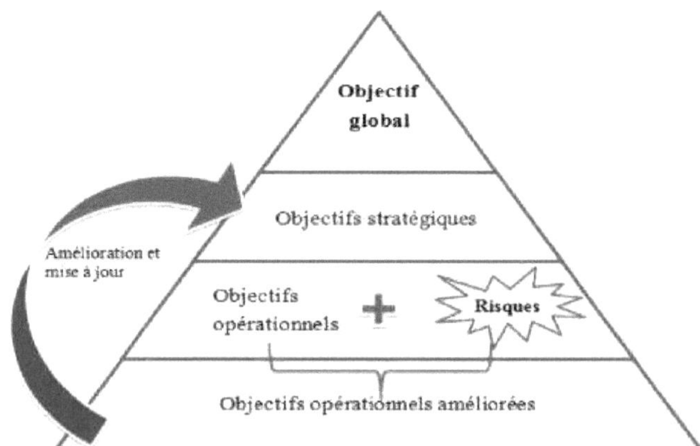

Figure 7.1 Model for validating strategic objectives via operational objectives

To feed our supply chain dashboard, we will suggest global indicators for the axes of the balanced scorecard that meet the defined strategic objectives. These indicators will be chosen on the basis of our bibliographic synthesis carried out in the framework of our article (Naciri et al, 2014b) and the level 1 indicators proposed by the SCOR model.

7.3 Review of the literature on global supply chain indicators:

7.3.1 SCOR Level 1 indicators

Level 1 of the SCOR model is based on an analysis of five key management processes:

(1) the processes that balance total demand and supply in order to develop a plan of action that best meets supply, production and delivery [Plan],

(2) processes that provide goods and services to meet expected or actual demand [Source],

(3) the processes that transform the product in its final stage to meet the intended or actual demand [Make],

(4) processes that provide finished goods and services to meet expected or actual demand, typically including order management, transport management and distribution management [Deliver],

(5) processes associated with returns or products returned after receipt for any reason. These processes extend to post-delivery customer support [Return] (SCC, 2011).

The five management cycle processes refer, firstly, to two categories of costs related to customer relations and internal operations. The latter are further subdivided into three and two main dimensions respectively (Stewart, 1996; Pittiglio et al. 1999) (Table 7.1).

Table 7.1 Performance attributes and SCOR level 1 metrics (Morana, 2008)

Métriques SCOR niveau 1	Attribut de performance				
	Orientation client			Orientation interne	
	Fiabilité	Réactivité	Flexibilité	Coûts	Actifs
Taux de livraison Parfaite	✓				
Délai d'exécution d'une commande		✓			
Adaptabilité de la Supply Chain à la hausse			✓		
Flexibilité de la Supply Chain à la hausse			✓		
Adaptabilité de la Supply Chain à la baisse			✓		
Coût total de la Supply Chain				✓	
Coût des produits Vendus				✓	
Durée de cycle d'exploitation					✓
Rotation des actifs de la Supply Chain					✓
Rotation du fonds de roulement					✓

SCOR Level 1 metrics	Performance attribute	

	Customer orientation			Oriental on internally		
	Reliability	Reactivity	Flexibility	Costs ▼	Assets	
Delivery rate Perfect	*J*					
Time limit for the execution of an order		*J*				
Supply chain adaptability to the rise			*J*			
Supply Chain Flexibility on the rise			*J*			
Adaptability of the supply chain to the downturn			*J*			
Total supply chain costs				*J*		
Cost of products sold				*J*		
Operating cycle time					*J*	
Supply Chain Asset Turnover					*J*	
Turnover of working capital					*J*	

The first category of costs related to customer relations can be broken down into three **elements, namely**

reliability, which is the ability to deliver the right product, in the right place, on time, in the right packaging, in the right quantity, with the right documentation and to the right customer;
- *The* aim *of* the project is to provide the products to each customer with speed and ;

- *flexibility* to respond to various environmental changes.

The second category of costs, those related to internal operations, are the financial reflection of the operationalisation of the supply chain and the efficiency of asset management.

In addition, SCOR takes into account the global environmental indicators of the GreenSCOR version (SCC, 2011). These indicators are :

Total carbon footprint of the supply chain

The sum of the carbon equivalent emissions associated with the supply chain.

Total environmental footprint of the supply chain

The sum of air, liquid and solid waste emissions associated with the supply chain.

7.3.2 Global indicators citis in the literature

In 2002, experts from twenty-nine U.S. companies and government agencies gathered in a focus group organised by The Logistics Institute (TLI) at the Georgia Institute of Technology, defined seven key supply chain indicators:

> the proportion of correct orders ;

> the percentage of deliveries made on time ;

> the degree of accuracy of the inventory ;

> the level of transport costs;

> the stock turnover rate ;

> adaptation to sales cycles;

> and the accuracy rate of purchase orders.

Their objective is to place themselves in a benchmarking logic by equipping themselves, thanks to these seven indicators, with a measurement system that makes it possible to compare logistics practices from one company to another or from one business sector to another. According to these experts, this measurement system is a simple but operational modelling of the link between the efficiency of the logistics function and the economic performance of the company.

The classification of supply chain performance indicators was the subject of a study by Taylor (2003) who classified them into four categories:

> ***Time measurements***: including among others order cycle time, product development cycle time, on-time delivery.

> ***Cost measures***: including raw material costs, payroll, maintenance, returns of defective products, transport, storage, and infrastructure management, sales growth, among others.

> ***Efficiency measures***: relate to the utilisation rate of a chain asset such as the utilisation rates of warehousing centres, the rate of production capacity utilisation, and the rate of capital utilisation.

> ***Service quality measures***: such as rates of on-time deliveries, satisfied orders, factory returns, customer complaints, and customers placing new orders.

For an overall optimisation of the supply chain, a balance between financial and non-financial metrics has to be found, be it at strategic, tactical or operational level. Also, being able to make the right measurements in all functions of the chain would allow to better understand it and thus to be able to improve it where the needs are felt. In this sense, Gunasekaran et al, (2004) identify the important indicators for which they develop metrics (Table 7.2).

Table 7.2 The six indicators according to (Gunasekaran et al, 2004) (Naciri et al, 2014b)

Metrics	Indicators	Definitions
Planning orders	Order lead time	The total order cycle is the time from receipt of the order to delivery of the product to the customer.
Evaluation of suppliers	Evaluation of suppliers	This evaluation has often been based on price variations and delivery times. The competition between suppliers was based on price, ignoring other equally important aspects such as quality, responsiveness, availability and customer satisfaction.
At the production level	production capacity	Its role is important as it determines the levels of activity throughout the chain. It directly influences the speed of response to orders (chain reactivity) and the cycle time of a product in the chain.
Evaluation of deliveries	Evaluation of deliveries	This evaluation determines to a large extent whether the customer is satisfied or not, and thus the competitiveness of the chain.
Quality of service assessment	Flexibility	the ability of the supply chain to respond favourably to individual customer requests. Flexibility can be measured by the cycle time of product development and the set-up times of machines or tools.
	Response time to customer requests	for example on the status of their orders.
Evaluation of the costs of logistics	Evaluation of the costs of logistics	It is a revaluation of all costs related to logistics. It is a very important financial indicator; financial flows have a great influence on product flows. One of its indicators is the measurement of the cost of risks undertaken by the chain.

Based on the work of (Gunasekaran et al, 2004; SCC, 2011; Gunasekaran et al, 2001; Barut et al, 2002; Sahin and Robinson, 2005; Wu and Song, 2005, De Toni and Nassimbeni, 2001), France-Anne (France-Anne, 2007) proposes a synthesis of indicators often cited in the literature to measure the performance of a supply chain (Figure 7.2)

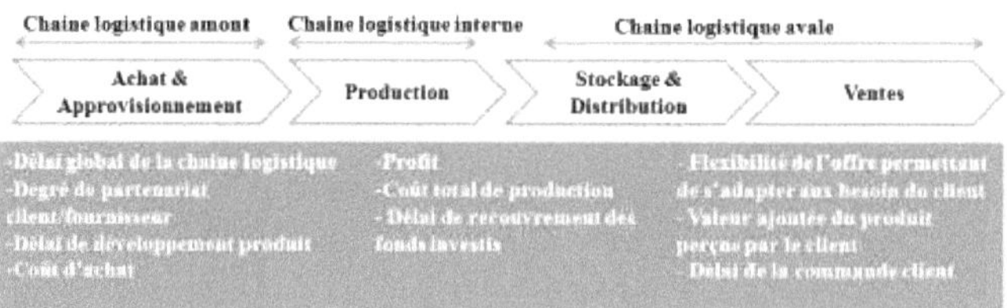

Figure 7.2 Main global indicators of company *performance* (France-Anne, 2007).

Julien Francois (Francois et al, 2007) identifies three global indicators of a supply chain:

- *The degree of partnership*

Different degrees of understanding between partners in a supply chain can be demonstrated. Lauras et al (2003) distinguish between 'communication', 'coordination', 'cooperation' and 'collaboration'. These different degrees of partnership depend on two factors:

- the type of information or processing (solving part or all of a problem) shared by the partners,

- the agreement to exchange or share this information between the two partners.

> *Cost reduction*

Reducing costs throughout the supply chain is one of the main priorities of supply chains. Reducing overall costs allows the price of finished products to be reduced and thus to seek new market share, and to free up profits for future supply chain investments.

> *Reduction of delivery time*

a company must react quickly to market changes in order to benefit from them.

In terms of environmental compliance, the Global Reporting Initiative (GRI) (GRI, 2013) proposes a benchmark of environmental indicators in several areas (materials, energy, water, biodiversity, emissions, effluents and waste, products and services, compliance with legislation, transport and general).

7.3.3 Summary of supply chain indicators cited in the literature

Following the state of the art on global supply chain indicators, we can classify them according to the three TBP performance axes, which are: Financial, customer, and internal process. However, we notice an absence of indicators for the organisational learning axis. Therefore, we resort to our bank of indicators designed during the elaboration of our questionnaire. We propose to summarise these indicators in Table 7.3.

Table 7.3 Summary of global indicators according to the TBP (Naciri et al, 2014b).

Axes	Indicators
Financial	Evaluation of the costs of logistics Cost reduction
	Purchase cost
	Total production cost
	Profit
	Cost of products sold
	Sales growth

	the level of transport costs;
	Total supply chain costs
Customer	Evaluation of deliveries Delivery rate Perfect Time limit for the execution of an order Flexibility The percentage of deliveries made on time
	The proportion of correct orders Supply Chain Adaptability down Supply Chain Adaptability up Supply Chain Flexibility up Satisfaction rate Rate of on-time delivery Customer complaints Reduction of the delivery time Response time to customer requests
Internal processes	Control latency Evaluation of suppliers Production capacity Utilization rate of storage centres Rate of production capacity used rate of capital used rate of accuracy of purchase orders Cycle time of an order Development cycle time Degree of customer/supplier partnership Overall supply chain lead time Timeframe for recovery of invested funds Proportion of correct orders ; Percentage of deliveries made on time ; Degree of accuracy of the inventory ; Stock turnover rate ; Adaptation to sales cycles; Order form accuracy rate Operating cycle time Supply Chain Asset Turnover Turnover of working capital
Focus learning organizational	Employee satisfaction index Absenteeism rate Changes in educational levels Rate of implementation of the training plan Progress bonus
Social and community security	Total carbon footprint of the supply chain Total environmental footprint of the supply chain GRI Indicators

Based on this state of the art on global indicators of a supply chain which has been published (Naciri et al, 2014b), we will try to adapt them to our case study by following phase 4 of our methodology.

7.4 Phase 4 application

7.4.1 Grouping of operational objectives

The ambition here is to group the selected operational objectives for each process against the four core areas: financial, customer, internal processes and organisational learning (Table 7.4).

Table 7.4 Operational objectives according to TDBP axes

Axis	Operational objectives
Financial	Controlling process costs
	Controlling the costs of non-quality
Customer focus	Respect for delivery times
	Reliability of delivery
Internal process axis	Controlling and respecting delivery preparation times
	Respecting order processing times
	Adhere to the conditioning programme
	Increase machine availability
	Controlling product quality
	Controlling the stock situation
	Have a safety stock
	Selection of suppliers
	Reduce product deviations
Organisational learning axis	Providing continuous training

We recall that these operational objectives are determined from the company's strategy (top-down approach) and the risks incurred in the processes.

7.4.2 Link between strategic and operational objectives

We will now revise the strategic objectives with the operational ones (buttom-up approach) (Table 7.5). In bold are the objectives added to validate this approach.

Table 7.5 Linkage of strategic objectives to operational objectives

Axes	Improved operational objectives	Revised strategic objectives
Financial	Controlling process costs	Controlling supply chain costs
Customer focus	Reduce customer complaints	Ensure customer satisfaction in terms of quality and **delivery reliability.**
	Respect for delivery times	
Axis Internal supply chain	Controlling the costs of non-quality	**Operational efficiency in terms of cost-quality-delay and optimal stock control: operational productivity**
	Controlling and respecting delivery preparation times	
	Respecting order processing times	
	Adhere to the conditioning programme	
	Increase machine availability	
	Controlling the stock situation	
	Have a safety stock	
	Controlling product quality	
	Reduce product deviations	
	Selection of suppliers	**Evaluation of suppliers**
Organisational learning axis	Providing continuous training	Developing human resources
Social and societal responsibility	-	Adopt a safety, environment and security policy

Interpretation :

The added strategic objectives are:

-S Delivery reliability: We have included the notion of delivery reliability in the strategic objectives of the supply chain, which is one of the main performance objectives and meets operational needs.

J Supplier evaluation: We have added this strategic objective to the operational objective 'supplier selection'. Thus, the selection of suppliers becomes a strategic decision that has a crucial impact on the overall performance of the company (Naciri et al, 2013), this decision aims to create and maintain a network of reliable and efficient suppliers needed by the principal to meet the competitive challenges (Aguezzoul et al, 2009). In addition, the various works in the field of supplier selection show that the QCD triptych (quality, cost, delivery) remains the most used criterion in this process (Naciri et al,

2013).

J Operational efficiency: The diagnosis of the supply chain by the SCOR model, led to the determination of the deficiencies altering the proper functioning of the

internal processes. This has an impact on the performance of the supply chain. Therefore, new targets have been set to eliminate these deficiencies.

To take these objectives into account at the strategic level, we have integrated the strategic objective *'Operational efficiency'.*

However, in order to maintain the coherence of the performance axes, we have dissociated the "quality" objective from the "quality, safety and environment" objective. In fact, product quality is a key requirement that will be addressed in the internal processes. The safety and environment aspect concerns the social and societal safety axis.

7.4.3 Establishing the causal chain linking the five perspectives

The strategy map provides the basis for structuring the balanced scorecard, which is the essential element of the strategic management system. It is the central point of this system, where it is placed as a common and comprehensible point of reference for the whole person1 of an organisation (Kaplan and Norton, 2003). It expresses the strategic assumptions and defines the causal relationships between the selected outcome measures and the determinants of performance (Alimazighi, 2004). Figure 7.3 illustrates this causal chain.

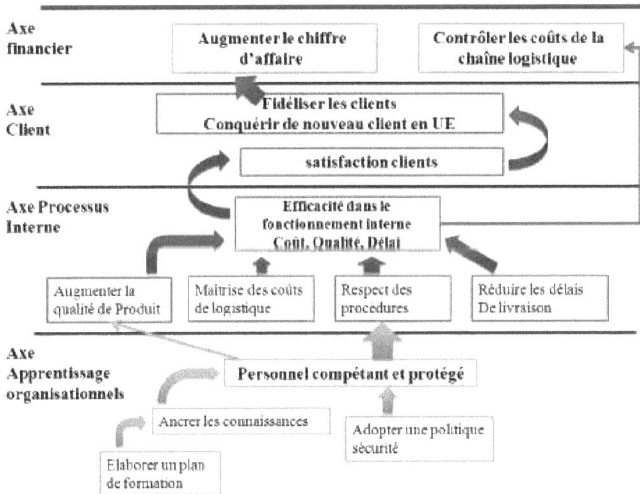

Figure 7.3 Strategic map of the station

7.4.4 Selection of relevant indicators by perspective

Performance management is based on the choice of key performance indicators in line with the objectives in order to build an adapted logistics dashboard. To do this, we create the link between the strategic objective and the indicators, based on the synthesis of the global indicators defined according to the Balanced Scorecard axes (Table 7.6).

Table 7.6 Overall indicators by perspective

Axes	Strategic objectives	Overall indicators
Financial	Controlling supply chain costs	Total supply chain costs
Customer focus	Ensure customer satisfaction in terms of quality and **reliability of delivery**	Customer satisfaction rate Perfect delivery rate
Internal process axis	**Operational efficiency in terms of cost-quality-delay and optimal stock control**	Operating cycle time Rate Internal environment Stock rotation
	Evaluation of suppliers	Evaluation of suppliers
Focus learning organizational	Developing human resources	Rate of implementation of the training plan
Social and societal responsibility	Adopt a policy, safety and environment	Accident frequency rate Total environmental footprint of the supply chain

Finally, we summarise our work in the form of a dashboard (Table 7.7) which represents the proposed performance indicators classified according to the five performance axes.

Thus, we present the calculation of these indicators, the target performance and the action plan.

Axes	Overall indicators	Calculation mode	Performance threshold	Action plan
Financial	Total supply chain costs	Supply cost + packaging cost + delivery cost	-	Identification and determination of sources of waste Optimal chain management
Customer focus	Customer satisfaction rate	1-(unsatisfied customers / total number of customers)	95%	Respecting the delivery Meet customer requirements Control product quality
	Perfect delivery rate	Total perfect deliveries / total number of deliveries	95%	Minimise order turnaround time Identification of sources affecting the operating cycle
Internal axis	Operating cycle time	the period between storage of the raw materials and payment of the invoice by the customer.	-	Reduce non-value added tasks
	Stock rotation	average stock for the period / average consumption for the period	-	Ensuring the availability of raw materials
	Rate Internal environment	Number of non-compliant tests / Total number of tests	<2%	A cleaning and disinfection check and sanitation in accordance with the instructions
	Evaluation of suppliers	Assigns a score from 1 to 5 based on price, quality, timeliness, responsiveness, and availability	-	Search for new suppliers
Organisational learning axis	Rate of implementation of the training plan	The training courses carried out / the training courses planned	95%	Follow continuous training
Social and societal responsibility	Accident frequency rate	Number of accidents* 10^6 /number of hours worked	-	Protecting staff
	Total environmental footprint of the supply chain	The sum of fair, liquid and solid waste emissions associated with the supply chain	-	Reducing emissions

Table 7.7 Overall Scorecard

116

7.5 Conclusion :

The purpose of this chapter was to design a global supply chain scorecard. For this purpose we drew on the TBP philosophy and our thinking about the determination of strategic objectives.

The starting point of the implementation is the strategic orientations of the company. These were enhanced and updated by the operational objectives of the supply chain processes.

To feed our chain dashboard, we have expanded our sources of indicators measuring the achievement of predefined objectives. Indeed, we used level 1 indicators from the SCOR and Green SCOR version 11 model, but due to the lack of indicators on organisational learning and staff safety and the low percentage of use of some indicators in practice (SCC, 2011), we built a bank of global supply chain indicators from our literature review (Naciri et al, 2014b).

Finally, we were able to build a global dashboard, which provides information on the achievement of the company's strategic objectives using relevant indicators capable of measuring the performance of the different dimensions of logistics performance.

General conclusion and future prospects

Our work has focused on measuring the performance of a supply chain through a dashboard. In order to design an efficient dashboard, we have succeeded in proposing a robust and generic methodology integrating the process approach recommended by the SCOR model and the Balanced Scorecard.

Starting from the performance dimensions of the Balanced Scorecard, our reflection consists in steering the internal processes that make up the supply chain to achieve a global steering of the chain. This reflection is translated into a methodology that covers all the elements required to establish an efficient performance indicator system and is based on four phases:

- ♦♦ Phase 1: Description of the field of study :

This phase consists of presenting :

- the company's integrated supply chain ;
- characteristics of the company :
- ■ general knowledge;
- ■ mission - vision - strategy ;
- ■ key processes.
- ♦♦ Phase 2: Global analysis and modelling

This phase consists of :

- analyse the company's inbound, internal and outbound logistics;
- performing a global supply chain modelling ;
- Conduct a SWOT analysis to highlight the company's strategic direction.
- ♦♦ Phase 3: Designing a process dashboard

We have proposed a general and rigorous approach for the development of a dashboard for the main processes based on the following steps:

> Step 1: Analysis of the functioning of the process ;

> Step 2: Risk identification and target setting;

> Step 3: Choice of indicators ;

> Step 4: Development of a process dashboard.

Knowing that the choice of really relevant indicators is the key to the success of any management project, we carried out an empirical study by questionnaire. This questionnaire survey was carried out among 266 industrial companies, with the aim of gathering the performance indicators considered useful for the main processes of the supply chain.

The bank of indicators obtained is used for a relevant choice of indicators in this phase.

♦♦♦ Phase 4: Design of a supply chain dashboard

To achieve this objective, we have drawn on the Balanced Scorecard approach to global performance measurement. Being "top-down", this approach is based on the assignment of performance indicators to the key success factors derived from the strategic objectives.

In order to converge the expectations/needs of the leaders with those of the actors in the field, we proposed to carry out a review and validation of the strategic objectives by those who are operational. This update seemed crucial to us in order to take into account operational constraints in the formalisation of the strategy.

To validate our methodology, we applied it in a small and medium-sized agri-food company wishing to initiate a project to develop a dashboard for the management of its supply chain.

To this end, we have succeeded in :

J Analysing and modelling the supply chain ;

J Study the company's strategic orientations;

J To study the malfunctions and risks within the processes affecting the smooth running of the supply chain;

J Designing dashboards by supply chain process ;

J Design a global supply chain dashboard that drives the company's strategy and takes into account operational constraints.

Finally, after summarising the results obtained, we will conclude this manuscript by proposing perspectives for future work in continuation of this research work.

We can consider some perspectives from an industrial point of view:

-   the computerisation and implementation of the dashboard production system, this is the stage where the choice of the software package and the platform to be used for the dashboard must be determined so that it is compatible with existing information systems.

From an academic point of view, two perspectives are possible to follow up on this work and thus complete the first results:

-   To examine the methodology for the management of integrated and collaborative supply chains. Considering the processes are the agents of the supply chain;

-   Designing supply chain risk dashboards ;
-   Steering a supply chain by introducing the concept of quality.

Bibliographic References

A

Agami, Saleh and Rasmy,2012. Supply Chain Performance Measurement Approaches: Review and Classification, Journal of Organizational Management Studies, IBIMA Publishing, 2012.

Aguezzoul, A. 2009 Selection and evaluation of the performance of logistics providers.

Amrani-Zouggar et al, 2007: A. Amrani-Zouggar, J. Francois, J.C. Deschamps and J.P. Bourrieres. Deschamps and J.P. Bourrieres. Multi-echelon production planning process with no time coordination. 4th IFAC Conference on Management and Control of Production and Logistics, 27-30 September 2007, Sibiu, Romania.

Anderson, S. W., Bechara, A., Damasio, H., Tranel, D., and Damasio, A. R. (1999). Impairment of social and moral behavior related to early damage in human prefrontal cortex. *Nat. Neurosci.* 2, 469-479. doi: 10.1038/12194

Ardoin, J.L., Michel, D., Schmidt, J. (1986), Le Controle de Gestion, 2eme ed. Publi-Union, Paris ASLOG (2006), Association Fran^aise pour la Logistique 2006

Atkinson, A., Epstein M. (2000), Measure for measure: realizing the power of the Balanced Scorecard, CMA Management, volume 74, n°7, pp. 22-28.

Azzouz Elhamma 2011; L'impact De La Strategie Sur Le Contenu Des Tableaux De Bord: Cas Des Entreprises Au Maroc Editions ICES; Revue Congolaise de Gestion 2011/2 Numero 14 | pages 57 a 77.

B

Barber, Brad M., Terrance Odean, and Ning Zhu. 2008. Do noise traders move markets? SSRN working paper.http://ssrn.com/abstract=869827.Barut et al, 2002;

Baumard P., Donada C., Ibert J. and Xuereb J.-M. (2003), "La collecte des données et la gestion de leurs sources", in R.-A. Thietart, Methodes de recherche en management, Dunod, Paris, pp. 224-256.

Beamon, B.M., 1998, "Supply chain design and analysis: Models and methods", International Journal of Production Economics, Vol. 55, No. 3, pp. 281-294.

Belin-Munier Christine. Logistics, supply chain and SCM in French-speaking management journals: what strategic dimension? XXIII eme conference de l'Association. Internationale de Management Strategique (AIMS), May 2014, Rennes, France. 26 p, Proceedings of the XXIIIrd conference of the International Association of Strategic Management (AIMS)

Berland N. (2009), Mesurer et piloter la performance, e-book, www.management.free.fr.

Berrah L., 1997. Une approche de revaluation de la performance industrielle: modele d'indicateur et techniques floues pour un pilotage reactif, These de doctorat de l'INP Grenoble.

Berrah, L, Cliville, V., 2007, Towards an Aggregation Performance Measurement System. model in a Supply Chain context, Computers in Industry, 58(7), September 2007, 709-719, ISSN: 0166-3615.

Bhagwat, R., Sharma, M. (2007a), Performance measurement of supply chain management: A balanced scorecard approach, Computers & Industrial Engineering, 53(1), pp. 43-62.

Bhagwat, R. An integrated BSC-AHP approach for supply chain management evaluation. Measuring Business Excellence ,v.11,n.3,p.57-68, 2007. http://dx.doi.org/ 10.1108 /136830 4071082 0755

Bhattacharya S, et al. (2014) Structural and functional insight into TAF1-TAF7, a subcomplex of transcription factor II D. *Proc Natl Acad Sci US A* 111(25):9103-8

Biteau R., Garreau A., Gavaud M., 1991, Dictionnaire des termes de gestion industrielle, Editions AFGI.

Bitton M., 1990, ECOGRAI: Methode de conception et d'implantation de systèmes de mesure de performance pour organisations industrielles - These de doctorat en automatique - Universite Bordeaux1

Bitton M., 1990, ECOGRAI: Methode de conception et d'implantation de systèmes de mesure de performance pour organisations industrielles - These de doctorat en automatique - Universite Bordeaux1.

Boix, D., and B. Feminier. 2003. Le tableau de bord facile. Paris Editions d'organisation, 274p.

Bonnefous C., 2001, La construction d'un systeme d'indicateurs pertinents et efficace, in Indicateurs de performance sous la direction de Chantal Bonnefous et Alain Courtois, Productique-Hermes, Paris.

Bonvoisin J., A. Lelah, F. Mathieux, D. Brissaud. 2011. Environmental Impact Assessment Model for Wireless Sensor Networks. In: *Glocalized Solutions for Sustainability in Manufacturing*, edited by J. Hesselbach and C. Herrmann, 124-129. Berlin: Springer.

Boudahri F, Z Sari, F Maliki, M Bennekrouf, Design and optimization of the supply chain of agri-foods: application distribution network of chicken meat, Communications, Computing and Control Applications (CCCA), 2013

Bouquin H., 2004, Le controle de gestion, 2eme edition, Collection Que sais je, Presses Universitaire de France, Paris.

Bourguignon L., Segard N., Deloze V., Sellami F. ,E M E Ry-Barbier A. et beyries s. 2001 - le gisement mousterien de "la folie" poitiers. dfs, service regionalde l'Archeologie de Poitou-Charentes

Bourne, M., Neely, A. D., Mills, J. F., Platts, K., and WILCOX, M., 2000, Designing,implementing and updating performance measurement systems, International Journal of Operations &Production Management , 20, (7), 754-771

Bowersox, D. J. Integrated Supply Chain Management: A Strategic Imperative. Conference of Council of Logistics Management, Chicago. 1997.

C

Camirenelli E., and Cantu A., (2006). Measuring the Value of the Supply Chain: A
Framework. Supply Chain PracticeVol. 8. No. 2. D. Brown and A. Sloan

Capar, S.G. (2002). Multi-element analysis of food by microwave digestion and inductively coupled plasma-atomic emission spectrometry. J. Food Compos. Anal. 2002;15:593-615.

Cohen, S., Gottlieb, B. H., Underwood, L. G., 2005, Social relationships and health, in Cohen, S., Underwwod, L. G., Gottlieb, B. H., eds, Social Support Measurement and Intervention, New York, Oxford University Press, 3-25

Colin J. (2005), "Le supply chain management existe-t-il reellement ?", Revue i'rancaise de gestion(3), n° 156, p. 135-149

Colin, J. and Pache, G. (1988), La logistique de distribution, Chotard et associes editeurs, Paris

Cooper M. C., Ellram L. M., Gardner J. T. and Hanks A. M. (1997), "Meshing Multiple Alliances", Journal of Business LogisticsVol. 18, pp. 67-89.

Cooper ct al, 1997: M.C. Cooper, D.M. Lambert and J.D. Pagh. Supply Chain Management: More Than a New Name for Logistics. The International Journal of Logistics Management8(1), pp 1-13, 1997.

Courty, 2003: P. Courty. Industrial issues and new scientific problems - From logistics to global logistics. Ecole d'ete d'automatique - Gestion de la Chaine Logistique. Session 24, September 2003, Grenoble, France.

D

Daniel Feisthammel Pierre Massot Fondamentaux Du Pilotage De La Performance , Brooch - Guide Published In 01 2005

Daum, J. 2005. French Tableau de Bord: Better than the Balanced Scorecard ?, Der Controlling Berater, No. 7, pp. 459-502.

Demarcheiso, 2010, http://www.demarcheiso17025.com

Distler,Fana RASOLOFO Conception et mise en rauvre d'un systeme de pilotage integrant la Responsabilité Sociale de 1 Entreprise : Une methode combinatoire, These en Sciences de Gestion,

Ecole Superieure de Management - Institut d'Administration des Entreprises, 2009.

Dixon J.R., Nanni A.J., Vollmann T.E. (1990), The new performance challenge: measuring manufacturing for world class competition, Dow-Jones-Irwin.

Doumeingts, B. Vallespir, M. Zanettin, D. Chen. - GIM: GRAI integrated methododology, a methodology for designing CIM systems. - Technical report for the IFAC/IFIP task force on architectures for integrating manufacturing activities and enterprises, version 1.0, GRAI/LAP, Bordeaux, France, May 1984

Dudek G. and Stadtler H., 2005, Negotiation-based collaborative planning between supply chains partners, European Journal of Operational Reseach, volume 163, Issue 3, p.668-687.

Dupont L. (2003), " Solutions Pratiques : logistique et Supply Chain, questions-reponses ", Tome 1, Editions WEKA.

Dupuy et al, 2004: C. Dupuy, V. Botta-Genoulaz and A. Guinet. Batch dispersion model to optimise traceability in food industry. Journal of Food Engineering, Special Issue on "Operational Research and Food Logistics", Vol. 70,Issue 3, pp 333-339, 2004

E

ECCLES, R. G., 1991, The performance measurement manifesto, Harvard Business Review, 131-137.

Epstein, M. Manzoni JF. 1998. Implementing Corporate Strategy: From Tableaux de Bord to Balanced Scorecards, European Management Journal, Vol 16, N°2 190-203.

EVALOG 2006: GLOBAL EVALOG frame of reference, 2006. Available online http://www.galia.com.

F

Fernandez, A. 1999. Les nouveaux tableaux de bord pour piloter l'entreprise. Paris: Editions d'organisation, 347p.

Fernandez, A. 2005. L'essentiel du tableau de bord, Edition d'organisation.

Fernandez, A. 2007. Les nouveaux tableaux de bord des managers, Edition d'organisation.

Florence. Gillet - Goinard, L. Maimi, Toute la fonction production, L'Usine Nouvelle/Dunod, 2007

France-Anne Gruat La Forme-Chretien, Referentiel d'évaluation de la performance d'une chaine logistique Application a une entreprise de l'ameublement 2007.

Francois et al, 2005: J. Francois, J.C. Deschamps, G. Fontan and J.P. Bourrieres. Macro-planning models for the management of logistics chains: a case study. Revue e-STA - www.see.asso.fr, 2005.

G

Geary, S. and Zonnenberg, J.P. (2000), "What it means to be best in class", Supply Chain Management Review ,Vol.4No.3,pp.43-8.

Gelinas, D., (2002), Le suivi intensif en equiped dans la communauteet et la readaptation psychosociale font-ils bon menage, Speech delivered in Montreal at the colloquium Suivi intensif enequipe dans la communaute: Fideliteau modele ACT etstrategie d'implantation organised by the Association des hopitaux du Quebec on 21 November 2002.

Gilmour P. A strategic audit framwork to improve Supply Chainperformance. Journal of business & industrial marketing Vol 14 No.5/6. pp.355-363. 1999.

GRI, Indicators & Protocols: Environment, version 4.0, 2013, 29 p.

Griffis, S.E., Goldsby, T.J., Cooper, M., Closs, D.J. (2007). Aligning logistics performance measures to the information needs of the firm, Journal of Business Logistics, 28(2), pp. 3556.

Guerny, J.; Guiriec, J. C.; Lavergne, J. (1990): Principes et mise en place du Tableau de Bord de Gestion, 6th Edition, Paris 1990.

Gunasekaran, A 2004, "Supply chain management: Theory and applications", European Journal of Operational Research, Editorial, Vol. 159, pp. 265-268.

Gunasekaran, A. and Kobu, B. (2007), Performance measures and metrics in logistics and supply chain management: a review of recent literature (1995-2004) for research and applications, International Journal of Production Research, 45(12), pp. 2819-40.

Gunasekaran, A., Patel, C. Tirtiroglu, E. (2001), Performance measures and metrics in a supply chain environment, International Journal of Operations & Production Management, 21 (1-2), pp. 71-87.

H

H. Zhao, N. Li, 2015. A statistical framework to predict functional non-coding regions in the human genome through integrated analysis of annotation data. Scientific Reports, 5: 10576.

Handfield Robert B., Straight Samuel (2004) The Supply Chain Maturity Model, Supply Chain Resource Consortium, North Carolina State University

Hendricks, K.B. and Singhal, V.R., 2005. An Empirical Analysis of the Effect of Supply Chain Disruptions on Long-Run Stock Price Performance and Equity Risk of the Firm. Productionand Operations Management, 14(1), pp.35-52.

Hendrik Reefke Mattia Trocchi , (2013), "Balanced scorecard for sustainable supply chains: design anddevelopment guidelines", International Journal of Productivity and Performance Management, Vol. 62 Iss 8pp. 805 - 826

Hockerts, Kai (2001): Corporate Sustainability Management - Towards Controlling Corporate Ecological and Social Sustainability. In: Proceedings of Greening of Industry Network Conference, 21-24 January 2001, Bangkok.

HolaBa B, Christian P Models for optimization and performance evaluation of biomass supply chains: An Operations Research perspective, Renewable Energy, Volume 87, Part 2, March 2016, Pages 977-989

Hollnagel, E., Woods, D., Leveson, N. 2006. Resilience engineering: Concepts and precepts. Aldershot, UK: Ashgate. 410 p.

Horvath, A. (2003), Energy-related Emissions from Telework. *Environmental Science & Technology*, 37(16), pp. 3467-3475,

Humez, 2008 *On the Dot: The Speck That Changed the World* by Alexander, Back Matter:" it. See, for example, the exuberant hypothetical recon- structions in Zacharie Mayani's *The Etruscans Begin to Speak*. 166 China

I

Ibn El Farouk I, Jawab F, Talbi A, Development Of A Set Of Indicators To Manage Medicines Supply Chain In Moroccan Public Hospital,Application Of The SCOR Model." Ijed: -161 International Journal Of E-Business And Developpement, ISSN Print No.Is: 2225-7411, ISSN Online No. Is: 2226-7336. Vol. 3 - N°3 - August2013.

J

Jaouhar Mahmoudi , Lamothe J., and Thierry C., A simulation model for customer-supplier cooperation in the telecom supply chain, International Journal of Business Performance Management, 2006 .

Juglaret. F Indicators and dashboards for risk prevention in occupational health and safety. Gestion et management. Ecole Nationale Superieure des Mines de Paris, 2012.

K

Kaplan, R.S. 1996, "Using the balanced scorecard as a strategic management system", Harvard business review, vol. 74, no. 1, pp. 75-85.

Kaplan, R.S. and Norton, D.P. 1992, "The balanced scorecard--measures that drive performance", Harvard business review, vol. 70, no. 1, pp. 71-79.

Kaplan, R. 1993, "Putting the balanced scorecard to work", Harvard business review, vol. 71, no. 5, pp. 134-144.

Kaplan, R.S. and Norton, D.P. 1996, "Linking the balanced scorecard to strategy", California management review, , no. 1, pp. 53-79.

Kaplan, R.S. and Norton, D.P. 2001, "Transforming the balanced scorecard from performance measurement to strategic management: Part I", Accounting Horizons, vol. 15, no. 1, pp. 87-104.

Kaplan, R.S. and Norton, D.P . Commentutiliser le tableaudebrospectif", Edition d'organisation2003

Kearney A. T. (1994), "Management approach to Supply Chain integration" ,Report to the members of the A.T. research team. Kearney.

Kpi 2015 referential on KPIs.

L

L'Houssaine (2013) performance et risques dans chaines logistique vulnerable : Analyse dans un contexte industriel marocain, 6eme edition du Colloque International LOGISTIQUA L'Ecole Nationale des sciences appliques de Tanger. 30 & 31 May 2013.

La Londe B.J. Evolution of the Integrated Logistics Concept. In ROBESON J.F., Copacino W.C., The Logistics Handbook, Free Press Edition. 1994.

Lambert and Cooper, 2000: D.M. Lambert and M.C. Cooper. Issues in Supply Chain Management. Industrial Marketing Management, 2000, 29, pp 65-83.

Langroodi R A system dynamics modeling approach for a multi-level, multi-product, multi-region supply chain under demand uncertainty, Expert Systems with Applications Volume 51, 1 June 2016, Pages 231-244.

Lauras et al, 2003: M. Lauras, N. Parrod, O. Telle, J. Lamothe and C. Thierry. Referentiel de l'entente industrielle: Trois approches dans le domaine de la gestion des chaines logistiques. 5eme International Congress of Industrial Engineering: Industrial Engineering and Global Challenges, Quebec City, Canada, October 2003.

Lavastre , O. Gunasekaran, A & Spalanzani, A. 2012 supply chain risk management in french companies decision support systems, 52(4), pp.828-838

Laverty J., Demeestere R. (1990), Les nouvelles regles du contrôle de gestion industrielle, Dunod, Paris.

Lavina, Y., Audit de la maintenance, Les Editions d'organisation, Paris, France, 1994.

Lee and Billington, 1993: H.L. Lee and C. Billington. Material management in decentralized supply chain. Operation Research, Vol 41, No 5, 1993.

Levi Simchi-, (2003) Production and Distribution Lot Sizing in a Two Stage Supply Chain. **IIE Transactions**, 35, pp. 1065-1075.

Lhoussaine Ouabouch and Mostapha Amri, "The performance of supply chains in the face of multiple upstream, internal and downstream disruptive incidents. Results of an empirical study in the Moroccan industrial sector", Question(s) de management 2014/1 (n° 5), p. 73-83.DOI 10.3917/qdm.141.0073

Likert, R. 1932 - A technique for the measurement of attitudes. Archives of psychology 140: Harvard Business Review36 (4):37-66.

Loning H., Pesqueux Y. (1998) Controle de gestion, Dunod

Lorino, P. 1997, 2001. Methodes et pratiques de la performance, Editions d'Organisation

M

Malo J. L. (2000), "Tableaux de bord" in Encyclopedie de Comptabilite, Controle de Gestion et Audit, Economica, pp. 1133-1144.

Malo J. L. (1992), "Tableaux de bord", in Encyclopedie du Management, Tome 2, vuibert,

Maria, R., 2012. Empirical study on the indicators of sustainable performance-the sustainability balanced scorecard, effect of strategic organizational change. Amfiteatru Econ. 32, 451e469

Mauchand, M., 2007, Modelisation pour la simulation de chaines de production de valeur en entreprise industrielle comme outil d'aide a la décision en phase de conception/ industrialisation Phd. Thesis Ecole Centrale Nantes 2007.

Mendoza C, Delmond M.H, Giraud F, Tableaux de bord et balanced scorecard, Editions Revues fiduciaires, 2005.

Michel, D., Gillet, G., Volovitch, M., Pessac, B., Calothy, G., and Brun, G. (1989). Expression of a novel gene encoding a 51.5 kD precursor protein is induced by different retroviral oncogenes in quail neuroretinal cells. Oncogene Res 4, 127-136.

Mikus, I. D. 2001 Fundraising for SME's, International Entrepreneur Conference, Paper and Proceedings, Nottingham, 2001.

Miller T., 2001, Hierarchical operations and supply chain planning, Springer.

Morana Joelle Le tableau de bord durable d'un systeme demutualisation des livraisons, Nantes Urban Logistics Conference, Jun 2008, Nantes, France

Morana Joelle, Jesus Gonzalez-Feliu. The sustainable scorecard of a delivery pooling system in the light of 21st century concerns. 2nd Nantes Urban Logistics Conference, Jun 2012, Nantes, France.

Morana, J. and Pinardi G. (2003), Elaboration d'un tableau de bord des couts logistiques de distribution, Revue Franchise de Gestion Industrialle, Vol. 22, n° 4, pp. 77-95

Morgan J. Integrated Supply Chains: How to make Them Work! Purchasing. May 1997 pp. 32-37. 1997.

Morley C. (2002), "Management d'un projet systeme d'information: principes, techniques, mise en œuvre et outils", 4th edition; Dunod, 360p.

Mouloua Z., and Oulamara A., (2007). Joint optimization of production and transportation in a multiple products, multiple customers supply chain. Paper submitted.

Moutaoakil H., JAMOULI H., Analysis and modeling of Moroccan citrus Supply chain based on Multi-Agent System and Performance Indicators, Responsive and Robust Planning, GOL, RABAT, MAROC, 2014

N

Naciri. O, Ayoub. A, Herrou. B, El Hammoumi (2012b) Realisation d'un Tableau de bord Global d'une chaine logistique" 2eme edition International Symposium on Security and Safety Complex Systems 2SCS 2012

Naciri. O, Ayoub. A, Herrou. B, El Hammoumi (2014a). M, Design Of An Evaluation System And Performance Management Of Supply Service: Case Study" International Journal of Scientific & Engineering Research Volume 5, Issue 1, January-2014, ISSN 2229-5518.

Naciri. O, Ayoub. A, Herrou. B, El Hammoumi (2014b). An approach based on balanced scorecard for the implementation of overall performance indicators of a supply chain. Naciri oumaima. International Journal of Scientific & Engineering Research Volume 5, Issue 7, July 2014, ISSN 2229-5518.

Naciri. O, Ayoub. A, Herrou. B, El Hammoumi (2015a). Investigation of the importance of performance indicators in the control of the supply chain of the Moroccan industrial sector Mohammed International Journal of Scientific & Engineering Research Volume 6, Issue 6, June 2015, ISSN 2229-5518.

Naciri. O, Ayoub. A, Herrou. B, El Hammoumi (2015b). Construction approach of a performance indicator system for controlling production process applied to a packing station for fruit and vegetables for export Line Spacing International Research Journal of Engineering and Technology (IRJET) E-ISSN: 2395 -0056 P-ISSN: 2395-0072.

Naciri. O, Ayoub. A, Herrou. B, El Hammoumi (2015c). Contribution to the management of supply chain processes: Case study of the procurement process of a Moroccan SME. Xeme Conference Internationale: Conception et Production Integrees, CPI 2015, 2-4 December 2015, Tanger - Morocco

Naciri. O, Ayoub. A, Herrou. B, El Hammoumi 2013 Proposal of performance indicators of supply service: Case study of a Moroccan SME". 6th edition of the International Symposium LOGISTIQUA National School of Applied Sciences of Tangier. 30 & 31 May 2013.

Naini, Aliahmadi and Jafari-Eskandari (2011) Using game theory and fuzzy MCDM to choose strategic orientation in uphold of private sector by approach balanced scorecard: Iranian industries, International Journal of Services, Economics and ManagementVolume 3, Issue 4. 2011

Narasimhan and Talluri 2009 perspectives on risk management in sypply chains. Journal of Operations management, 27(2), pp. 114-118.

Narasimhan, R., & Talluri, S. (2009). Perspectives on risk management in supply chains. Journal of Operations Management,27(2),114-118.

Newsted P., Huff S., & Munro M. (1998) 'Survey Instruments in IS' MISQ Discovery, December 1998,

Norreklit, H. 2000. The balance on the balanced scorecard - a critical analysis of some of its assumptions, Management Accounting Research, Vol 11, 65-88.

O

O'Brien, D.P. 2000. Business Measures for Safety Performance. Lewis Publishers, Washington, 118 p.

Okar Chafik , Zitouni Beidouri, Said Mssassi, Said Barrijal, A,. (2011) maturity model for SCPMS project: an empirical investigation in large sized Moroccan companies, International Journal of Computer Science Issues, vol. 8, n° 2, pp. 203-212, 2011.

Ouzizi et al, 2005: L. Ouzizi, M.C. Portmann, F. Vernadat. Decision support for supply chain planning using a semi-distributed control architecture. 6emeCongres International de Genie Industriel, 7-10 Juin 2005, Besan^on, France.

P

Pichot L. (2006), "Strategie de déploiement d'outils de pilotage de chaines logistiques: Apport de la classification", These de doctorat en Informatique et Systeme Cooperatifs, L'institut national des sciences appliques de Lyon.

Pichot L., Neubert G., Baptiste P. - "Customer segmentation in a supply chain environment"- International Conference on industrial engineering and Production Management IEPM'2003 - Porto (Portugal) -May 26-28, 2003 - ISBN 2-930294-13-02, proceedings on CDROM, 10 p.

Pinsonneault, Alain E Kraemer, Kenneth. Survey Research Methodology in Management Information Systems: As Assessment. Journal of Management Information Systems, Fall 1993.

Pirard, R., 2005, "Pulpwood plantations as carbon sinks in Indonesia: methodological challenge and impacts on livelihoods", Proceedings of the CIFOR Workshop "Carbon Sequestration and Sustainable Livelihoods", 16-17 February, Indonesia, pp. 74-91.

Pittiglio, Rabin, Todd, McGrath (1999), Supply chain: instructions for use. Les bonnes pratiques du supply chain management, Logistiques Magazine, dossier, June 1999.

Q

Quivy R., Van Campenhoudt L., Manuel de recherche en sciences sociales, Dunod, 1995

R

Rahanandeh Langroodi, A system dynamics modeling approach for a multi-level, multiproduct, multi-region supply chain under demand uncertainty, Expert Systems with Applications Volume 51, 1 June 2014, Pages 231-244

Rajesh Asija, Patel Jaimin, Sangeeta (2012), a novel approach: pulsatile drug delivery system; international research journal of pharmacy, 2012. issn 2230 -8407

Ravignon, P.L. Bescos La methode ABC/ABM: Piloter efficacement une PME Les editions d'organisation (1998)

Regragui Y, Mohammen A, Maturite de la chaine logistique, levier de performance des entreprises. 6th edition of the International Symposium LOGISTIQUA The National School of Applied Sciences of Tangier. 30 & 31 May 2013.

Reiter, R.J., Tan, D.X., Manchester, L.C., Qi, W. (2001a) Biochemical reactivity of melatonin with re-active oxygen and nitrogen species: a reviewof the evidence. Cell Biochem. Biophys, 34:237256

Ritchie, B., Brindley, C., 2007a. Supply chain risk management and performance - A guiding framework for future development. International Journal of Operations & Production Management 27, 303-322.

Robin, G., Debonis, S., Dornier, A., Cappello, G., Ebel, C., Wade, R. H., Thierry-Mieg, D., & Kozielski, F. (2005). Essential kinesins: characterization of Caenorhabditis elegans KLP- **15.** *Biochemistry, 44,* 6526-36

S

Sahin, F and Robinson, EP. 2005. Information sharing and coordination in make-to-order supply chains. Journal of Operations Management, 23(6): 579-598.

Samuel, H.H., Sunl, K.S. and Wang, G. (2004), "A review and analysis of supply chain operations reference (SCOR) model", Supply Chain Management: An International Journal, Vol. 9 No. 1, pp. 23-9.

Saulquin, J.Y., Maupetit, M. 2004. C. EVA, performance and banking evaluation. JoumAe de recherche CERMAT a La performance: de la mesure N l' action a. 15 January 2004.

Saulquin, J.Y., Schier, G. 2005. CSR as an obligation/opportunity to revisit the concept of performance? CongrEs Grefige Nancy 2005.

Selmer C. (2003), Concevoir le tableau de bord Outil de contrôle, de pilotage et d'aide a la decision, 2eme edition, Dunod, Paris, 289 p.

Sharma M., Bhagwat R. (2007), An integrated BSC-AHP approach for supply chain management evaluation, Measuring Business Excellence, Vol. 11, N°3, pp. 57-68.

Sicotte, C., Champagne, F., Contandriopoulos, A.P. 1999. Organizational performance of public health organizations. Ruptures 1999, 6(1): 34-46

Simatupang, T.M., Wright, A.C. and Sridharan, R. (2004), "Benchmarking supply chain collaboration: an empirical study", *Benchmarking: An International Journal*, Vol. 11 No. 5, (forthcoming).

Simchi-Levi, D., Kaminsky, P. & Simchi-Levi, E. Designing and Managing the Supply Chain: Concepts, Strategies, and Case Studies. 2nd Ed. New York: The McGraw-Hill Companies, 2003.

Stadtler and Kilger, 2000: H. Statler and C. Kilger. Supply Chain Management and Advanced Planning: concepts, models, software and case studies, Springer Verlag, 2000.

Stadtler, H., 2005. Supply chain management and advanced planning - basics, overview and challenges European Journal of Operations Research, 163,575-588

Steadtler H., Kllger C., 2001,An overview in supply chain management and advanced planning, Springer. Pirard, 2005.

Stevens G. (1989), "Integrating the Supply Chain", International Journal of Physical Distribution and Materials Management, vol. 19, No. 8, pp. 3-8.

Stewart G. (1996), Optimising your logistics costs with simple and effective modelling, Logistique & Management, Vol. 4, N°2, pp. 65-72.

Supizet J. (2002), " Total Balanced Scorecard, un pilotage aux instruments ", L'Informatique Professionnelle n° 209, pp. 15-20

Supply Chain Council, SCC, 2000 -2006-2011: Supply Chain Council. http://www.supply-chain.org.

T

Tan K.C.. 2001 A framework of supply chain management literature. European Journal of Purchasing and Supply Management 7, 2001, pp 39-48.

Tang and Nurmaya Musa 2010. Identifying risk issuesand research advancements in syplly chain risk management; International Journal Of Production Economics, 133(1), pp.25-34.

Tang, O. and Nurmaya Musa, S., 2011. Identifying risk issues and research advancements in supply chain risk management. International Journal of Production Economics, 133(1), pp.25 34.

Tayur et al, 1999: S. Tayur, R. Ganeshan, M. Magazine. Quantitative models for supply chain management. Kluwer Academic Publishers, 2000.

Tchernev N., 1997, Modelisation of the logistic process in flexible production systems. D. thesis in computer science, Universite Blaise Pascal, Clermont II.

Thierry and Bel, 2002: C. Thierry and G. Bel. Management of logistic chains in the aeronautical field: decision support tools for partnership improvement. Revue Fran^aise de Gestion Industrielle, 2002.

Tsalis, T.A., Nikolaou, I.E., Grigoroudis, E., 2013. A framework development to evaluate the needs of SMEs in order to adopt a sustainability-balanced scorecard. J. Integr. Env. Sci. 3e4, 179e197

V

Valentine, J.P., R.H. Magierowski, and C.R. Johnson 2007, 'Mechanisms of invasion: establishment, spread and persistence of introduced seaweed populations' , Botanica Marina, vol. 50, pp. 351-36

Valla A., V. Botta-Genoulaz, A. GuineT and F. RIANE, 2005 Business Process Improvement using Simulation: An industrial Application. International Conference on Industrial Engineering and Systems Management (IESM'05), 2005, May 16-19, Marrakech (Morocco, pp. 884-893,), Proceedings ISBN 2-9600532-1-4.

Van der Stede, W. A., S. M. Young, and X. C. Chen. 2005. Assessing the quality of evidence in empirical management accounting research: The case of survey studies.Accounting, Organizations and Society 30 (7 / 8): 655-684

Van Der, Vorst. & Beulens, A. (2002). Identifying sources of uncertainty to generate supply chain redesign strategies. International Journal of Physical Distribution& Logistics Management,32(6),409-430.

Verane Humez proposal for a decision support tool for order management in case of shortage: a performance-based approach, thesis L'Institut National Polytechnique 2008

Villarmois , O. 2001. The concept of performance and its measurement: a state of the art. Les cahiers de la recherche. CLAREE, UPRESA CNRS 8020.

Voyer P. (2000), Tableaux de bord de gestion et indicateurs de performance, 2eme edition, presses de l'Universite du Quebec, 446 p

Voyer, Pierre. 2002. Tableau de bord de gestion et indicateurs de performance. Quebec: Presse de l'universite du Quebec, 446p.

W

Wattky, G. Neubert, Process reengineering in the context oflogistics outsourcing, accepts an Int. J. of Information Technology and Management, Special Issue "Business ProcessManagement: from Engineering to Execution", 2006

Winkler H. & Kaluza B., (2006), "Integrated Performance and Risk Management in Supply Chains - Basics and Methods", W. Kersten and T. Blecker (Eds.), Managing Risks in Supply Chains - How to Build Reliable Collaboration in Logistics, Erich Schmidt Verlag, Berlin, pp. 19-36;

Wong, W.P. and Wong, K.Y. (2008), "A review on benchmarking of supply chain performance measures",Benchmarking: An International Journal, Vol. 15 No. 1, pp. 25-51.

Wreathall, J. 2009. Leading? Lagging? Whatever! Safety Science 47, 493-494.

X

Xian, Qiu and Zhang (2013), Activation of receptor for advanced glycation end products contributes to aortic remodeling and endothelial dysfunction in sinoaortic denervated rats Original Research Article *Atherosclerosis, Volume 229, Issue 2, August 2013, Pages 287-294*

Z

Ziegenbein, A., Nienhaus, J.: (2004) Coping with Supply Chain Risks on Strategic, Tactical and

Operational Level, In: Harvey, R.J., Geraldi, J.G., Adlbrecht, G. (Eds.), Proceedings of the Global Project and ManufacturingManagement Symposium, pp. 165-180, Siegen, May 2004

Zimmermann, K., S. Seuring (2009). Two case studies on developing, implementing and evaluating a Balanced Scorecard in distribution channel dyads, International Journal of Logistics: Research and Applications, Vol. 12, pp. 63-81.

Annex 1

Questionnaire

Survey on process performance indicators

procurement

Questionnaire for the Procurement Manager

Identification of performance indicators for the procurement process

Organisations only evaluate their performance through the implementation/acting of qualitative and quantitative, financial and non-financial indicators, whose role is to provide a diagnosis of the practices of the whole firm. The aim of this questionnaire is to identify the different potential indicators of the supply process according to the typology of each company.

Company :
Address:
The editor's profession:

1. What type of business are you in?

Large company SME SMI

2. What is your company's activity

		Important	Rather important	Not important
FINANCIAL ASPECTS				
1	- Decrease in purchase price compared to historical price			
2	- Evolution of purchase price/ market price			
3	- Purchase turnover			
4	- Service cost / purchase turnover managed by the service			
5	- Service cost/savings generated by the service.			
6	- Average value of an order			
7	- Average cost of placing an order			
8	- Purchase amount in Life Cycle Cost			
9	- Difference between forecasts and orders			
10	- Stock evolution			
11	- Evolution of material consumption			
12	- Increase in supplier payment times			
13	- Evolution of the Total Cost of Ownership			
14	- Financial investments in stock			
15	-Annual purchase value per supplier			
CUSTOMER FOCUS				
16	- Satisfaction rate			
17	- Cumulative number of days late / number of late deliveries			
18	- Actions influencing market share			
19	- Actions affecting customer loyalty			
	INTERNAL PROCESSES			
20	- Individual performance / objectives			
21	- No. of purchase requisitions processed (+ / headcount)			
22	- Average time to process a Purchase Request			
23	- Differences between quantities received and quantities ordered			
24	- No. of non-compliant batches / No. of batches received			
25	- No. of batches received within the deadline / No. of batches received - % of buyers using the Internet at least once a week			
26	- Participation in trade fairs			
27	- Order automation rate			
28	-Average time to complete an order			
29	- Number of orders to suppliers ^ Total number of orders			
30	- No. of EDI suppliers / total no. of suppliers			
31	- No. of active suppliers monitored			
32	- No. of managed framework contracts			
33	- No. of items managed under framework contract / No. of items managed			
34	- Application rate for framework contract			
35	- No. of suppliers under progress plan			
36	- Rate of active suppliers			
37	- Taxed supplier rates			
38	- No. of suppliers with framework contract			
39	- No. of suppliers involved upstream			
40	- No. of peer suppliers			
41	- Number of suppliers under Quality Assurance			
42	- No. of ongoing consultations (/geographical area)			
43	- No. of benchmarks engaged			

44	- Globalization approach			
45	- No. of Urgent Purchase Requests / No. of AD			
46	- No. of disputes			
47	- Stock rotation by product type			
48	- Rejection rate due to quality defects			
50	- Number of out-of-stock situations that caused interruptions in production			
51	- Number of order changes classed by cause			
52	- Number of orders received and in progress			
53	-Employee productivity and workload			
54	- Participation rate in Make or Buy			
55	- % of specifications drawn up with the purchasing department			
56	- No. of functional CDCs / No. of technical CDCs			
57	- Rate of studies or works completed on time			
58	- Reduction in the number of suppliers			
59	- No. of entries and exits from the panel			
60	- Geographic location of strategic suppliers			
61	- Supplier response rate			
	ORGANISATIONAL LEARNING FOCUS			
62	- of buyers in training seminars			
63	- No. of subscriptions to technical journals / databases			
64	- Existence of a knowledge capitalization tool			
65	- Turnover			
66	- Purchase turnover covered by the service / total purchase turnover			
67	- Absenteeism rate			
68	- Number of hours of training			
69	- Purchase turnover / headcount			
70	- Progress bonus			

Questionnaire

Survey on process performance indicators

Production

Questionnaire for the Production Manager

Identification of performance indicators for the production process

Organisations only evaluate their performance through the implementation/acting of qualitative and quantitative, financial and non-financial indicators, whose role is to provide a diagnosis of the practices of the whole firm. The aim of this questionnaire is to identify the different potential indicators of the production process according to the typology of each company.

Company :

Address:

The editor's profession:

1. What is the type of your business?

Large company | SME |SMI

2. What is your company's activity

		Important i Somewhat i Not i important i important		
FINANCIAL ASPECTS				
1	- Production cost vs last year vs budget			
2	- Production cost - Sales cost			
3	- Fixed production costs			
4	- Variable production costs			
5	- Average production cost for the period			
6	- Actual production cost - Standard production cost			
7	- Cost associated with increasing production by 1 unit			
8	- Costs associated with machine downtime			
9	- Standard cost of raw materials used in actual production - Actual cost of raw materials			
10	- Standard cost of hours worked - Actual cost of hours worked			
11	- Value of finished products - Total production cost			
12	- Total HR costs - total number of units produced			
13	- Total raw material cost ^ total number of units produced			
14	- Total production cost - total number of units produced			
15	- Cost of product defects due to raw material quality - Total cost of defects			
16	- Cost of damaged products due to staff errors - total cost of damaged products			
17	- Standard cost of raw materials used in production - Actual cost of raw materials			
18	- Cost of production capacity - Cost of used production capacity			
19	- Quality control costs - Production costs			
20	- Costs of internal and external non-conformities + costs of inspections + costs of prevention			
CUSTOMER FOCUS				
21	- Satisfaction rate			
22	- Cumulative number of days late / number of late deliveries			
23	- Number of customer orders per day (units / day) - Number of minutes worked per day (minutes / day)			
	INTERNAL PROCESSES			
24	- Margin by product category			
25	- Actual production - planned production			
26	- Average units produced per day			
27	- Number of defects produced due to raw material quality -total number of defects			
28	- Damaged products due to staff error - total number of damaged products			
29	- Value of production losses - Value produced			
30	- Raw material loss - total raw material consumption			

No.	Description			
31	- Standard quantity of raw material theoretically required for production - quantity actually used			
32	- Number of days the units are in the production process without being considered finished products			
33	- Number of units in the production process that are not considered finished products			
34	- Value of units in the production process			
35	- Time between the first planned Production Order and the delivery of the ordered Finished Product			
36	- Standard cycle time ^ actual cycle time			
37	- Actual cycle time ^ ideal cycle time (minimum cycle time)			
38	- Number of times the production time is less than the maximum time allowed to satisfy the customer's demand			
39	- Production capacity - actual production			
40	- Actual production ^ production capacity			
41	- Total volume that can be produced			
42	- Availability rate x Performance rate x Quality rate			
43	- Actual production time / Theoretical production time			
44	- Cycle time x Actual production/ Actual production time			
45	- (Actual output - Rejected output)/ Actual output			
46	- Conforming finished products ^ total production			
47	- Number of pieces produced during the production time ^ maximum production rate			
48	- Actual production rate ^ target production rate			
49	- Machinery and equipment capacity used in production ^ Total machinery and equipment capacity			
50	- Actual production time and % of total time			
51	- MTBF - Mean Time Between Failures = total (uptime - downtime) ^ number of failures			
52	- Frequency of failures			
53	- Downtime for corrective maintenance			
54	- Downtime for preventive maintenance			
55	- Loss of time due to return to production x productivity x price			
56	- Number of parts rejected in the production process and to be returned to production for rework			
57	- Amount needed to return to production ^ number of pieces returned to production			
58	- Finished products according to quality standards ^ Total production			
59	- Number of defects ^ Product size			
60	- Number of defects ^ number of units produced			
61	- Number of non-compliant products ^ Number of quality checks			
62	- Number of items requiring a return to the production process			
63	- Number of pieces that need to be returned to the production process during pre-runs			
64	- Number of defaults Number of opportunities per default x 1 Million			
65	- Number of finished production orders per hour ^ total number of production orders			
66	- Number of late finished Production Orders ^ total number of Production Orders			
67	- Number of finished production orders in advance ^ total number of production orders			
68	- Number of production **delays** due to supply disruptions raw material			
69	- Total number of production delays			
70	- Note of finished products ^ Total production			
	LEARNING AXIS ORG ANISATI CN NE L			
71	- Total value produced ^ Number of employees			
72	- Standard number of hours required to complete the production ^ actual number of hours required			
73	- Production stoppages due to lack of staff training / Total production stoppages			
74	- No. of subscriptions to technical journals / databases			
75	- Existence of a knowledge capitalization tool			
76	- Progress bonus			
77	- Absenteeism rate			
78	- Number of hours of training			

Survey on process performance indicators
delivery

Questionnaire

Questionnaire for the Delivery Manager

Identification of performance indicators for the delivery process

Organisations only evaluate their performance through the implementation/acting of qualitative and quantitative, financial and non-financial indicators, whose role is to provide a diagnosis of the practices of the whole firm. The aim of this questionnaire is to identify the different potential indicators of the delivery process according to the typology of each company.

Company :
Address:
The editor's profession:

1. What type of business are you in?

       Large company J SME SMI

2. What is your company's activity

		Important	Rather important	Not important
FINANCIAL ASPECTS				
1	Evolution of cost vs. budget			
2	Cost of sales = Opening stock + Purchases of goods - Closing stock			
3	Total transport cost			
4	Transport cost / transport value (at cost of sales)			
5	Own transport cost $^\wedge$ Total transport cost			
6	Transport cost under drafts $^\wedge$ Total transport cost			
7	Cost of rental or depreciation of trucks			
8	Shipping cost Product			
9	Payroll $^\wedge$ number of drivers			
10	Transport cost $^\wedge$ Cost of sales			
CUSTOMER FOCUS				
11	- Number of customer orders delivered per day per FTE			
12	Number of trucks			
13	Annual number of deliveries (or tons, volumes of books...)			
14	Number of km, number of hours of use, number of collection points per day			
15	Average number of stops per trip			
16	Number of hours in use $^\wedge$ Number of hours available during the same period			
17	Number of unladen km $^\wedge$ total number of km done during the same period			
18	Capacity used (m3)$^\wedge$ capacity available (m3) during the same period			
19	Delivery variances			
20	Number of order lines in dispute $^\wedge$ Total number of order lines delivered during the same period			
21	Amount of goods in dispute (at cost of sales) $^\wedge$ total value of goods transported (at cost of sales) during the same period			
22	Number of deliveries per hour $^\wedge$ Total number of deliveries during the same period			
23	Waiting time per trip, per truck			
24	The time associated with receiving, entering and validating a customer order			
25	times associated with loading product and generating shipment			
26	The average time associated with the selection and rating carriers for shipments			
27	% of orders delivered in full			
28	Order execution time			
29	The average time associated with the shipment of products			
	ORGANISATIONAL LEARNING FOCUS			
30	Absenteeism rate			
31	Number of hours of training			
32	Amount of drivers' salaries			
33	Number of drivers			

Interview guide

1. What do you think is the vision, strategy and mission of the company?
2. What do you think are the activities and/or steps that lead to the performance of your company?
Understanding of the company's goals and objectives (mission and strategy)
3. What are the company's activities in its sector?
4. What products and services are offered?
5. What are the key processes in your company?
6. What is the geographical distribution of sales?
7. Who is the company's customer base?
8. Who are the company's suppliers?
9. What is the competitive situation? Who are the competitors? What are the key success factors (KSF) of the sector and the company? What are its distinctive competences?
10. What software is used?
11. What is your information system regarding customer satisfaction?
12. What are your short, medium and long term goals?
13. What means do you take to ensure the success of your company? (Strategy: how to achieve mission and objectives, to gain an advantage over the competition)
14. Do you have a system in place to know how you are performing financially, customer-wise, process-wise and in terms of innovation and learning?
15. Are there performance indicators in the organisation? If so, what are they? How are they measured? How often are they produced? Who uses these indicators and what actions do these users take as a result of analysing this information?
16. Are there any performance indicators not currently produced that the organisation considers relevant?

Printed by Books on Demand GmbH, Norderstedt / Germany